Perspectives on Human-Animal Communication

Despite its inherent interdisciplinarity, the Communication discipline has remained an almost entirely anthropocentric enterprise. This book represents early and prominent forays into the subject of human-animal communication from a Communication Studies perspectives, an effort that brings a discipline too long defined by that fallacy of division, human or nonhuman, into conversation with animal studies, biosemiotics, and environmental communication, as well as other recent intellectual and activist movements for reconceptualizing relationships and interactions in the biosphere. This book is a much-needed point of entry for future scholarship on animal-human communication, as well as the whole range of communication possibilities among the more-than-human world. It offers a groundbreaking transformation of higher education by charting new directions for communication research, policy formation, and personal and professional practices involving animals.

Emily Plec is Professor of Communication Studies at Western Oregon University, US.

Routledge Studies in Rhetoric and Communication

Perspectives on Human-Animal Communication

Internatural Communication

Edited by Emily Plec

Routledge
Taylor & Francis Group

NEW YORK   LONDON

First published 2013
by Routledge
711 Third Avenue, New York, NY 10017

Simultaneously published in the UK
by Routledge
2 Park Square, Milton Park, Abingdon, Oxfordshire OX14 4RN

First issued in paperback 2014

*Routledge is an imprint of the Taylor and Francis Group,
an informa business*

Library of Congress Cataloging-in-Publication Data
  Perspectives on human-animal communication : internatural
communication / edited by Emily Plec.
    p. cm. — (Routledge studies in rhetoric and communication ; 12)
  Includes bibliographical references and index.
  1. Human-animal communication.   I. Plec, Emily, 1974–
  QL776.P47 2013
  591.3—dc23
  2012031660

ISBN 978-0-415-64005-3 (hbk)
ISBN 978-1-138-92187-0 (pbk)
ISBN 978-0-203-08293-5 (ebk)

Typeset in Sabon
by IBT Global.

For

Lila & Meisie
Rhombus, Rory, Porter, Tootie, Lou, and Feister
Charlemagne
Wil & Henry
Marley
Asti
Bronte, Toklas & Vita
Bodhi, Duke, Ginger, Minden, Babe, Bugs, & Ellie
Spot, MoDog, Nellie, Rolly, & Molly
Scooter, Zoe, Winkin, Pandora, S'okay, Zen, Moco, Star,
Smiley, Puppy, Mustache, Granny, Luna & Ruffles
and all the friendlies

and to our human loved ones, for their support and generosity.

And especially for our late friend Nick Trujillo, a wonderful
person, scholar, and companion to humans and animals alike,
who inspired this collection by affirming for a doubtful young
graduate student that there was a space in the discipline for a
paper on dogs.

Contents

PART II
Implication

PART III
Coherence

Figures

Preface

Over a decade ago, when I first began to consider human-animal communication from my disciplinary perspective, I found a small but supportive community of scholars who were interested in similar and related questions. For various reasons, we had come to communication studies and found a frustrating obsession with the animal-human dichotomy, often manifest in statements such as "What sets humans apart from other animals is their capacity to communicate using symbols" but also found in the common dismissal of human-animal relationships as insignificant interpersonal phenomena. We knew from our own experiences with animals that communication theories and methods could prove insightful, and we hypothesized that there was more to the communication relationship than ethologists or lay advocates of human-animal interaction might have already shown.

Many years after my first foray into scholarly examination of human-canine communication, I find myself back at the same point, seeking a communication discipline that is inclusive of all animals—indeed, of all life—and views the theoretical resources of our discipline as starting points for a greater understanding of how best to live together. I am grateful to the authors of this collection, as well as to my nonhuman and human teachers, for exploring these generous and insightful possibilities with me.

Emily Plec

Acknowledgments

This book is the collective effort of a committed group of scholars. Although dedicated to the animal companions who have exercised patience in teaching us about their worlds, this volume was made possible by innumerable other humans who put up with us and inspired our work, including our partners, children, mentors, advisors, colleagues and the many scholars of animal communication cited in these essays. Without such support, we rarely test the boundaries of our ways of thinking and theorizing. With it, the possibilities are as great as our imagination and effort.

I am also indebted to the anonymous reviewers of the proposal for this volume, whose cautious enthusiasm and advice helped to prepare me for the task of editing and warranted the publisher's faith in the project. II extend my deepest gratitude to the authors of the chapters in this volume, to Diane Huddleston for her editorial assistance, and to Liz Levine, Andrew Weckenmann and Michael Watters, whose professionalism and support helped to make this project a labor of love.

The editor and publisher would also like to thank the following for granting permission to reproduce material in this work:

Taylor & Francis for permission to reprint "Communicating Social Support to Grieving Clients: The Veterinarians' View" by Mary Pilgram. This chapter was originally published in *Death Studies*, September 2010, Vol. 34, 8. Reprinted by permission of Taylor & Francis Ltd, http://www.tandfonline.com.

Ethan Welty Photography for permission to reprint the photographs *Girls and Gorilla, Woodland Park Zoo, Seattle, WA* and *Sleeping Gorilla, Woodland Park Zoo, Seattle, WA.*

Tema Milstein for permission to reprint the photograph *Akenji Pounds on the Glass.*

Joy Harjo and the W.W. Norton Company for permission to reprint three lines from "Remember." Copyright © 1983 by Joy Harjo, from SHE HAD SOME HORSES by Joy Harjo. Used by permission of W.W. Norton & Company.

1 Perspectives on Human-Animal Communication
An Introduction

Emily Plec

"Does every intelligent creature have to do things of which we can see the point and show its intelligence in ways we can recognize?"

—Mary Midgley (qtd. in McReynolds 157)

"We stand in community with other animals by virtue of our communication with them."

—Douglas Anderson (190)

Many students of communication are drawn to the field, as I was, because of its inherent interdisciplinarity and because of its capacity to be inclusive of a wide range of perspectives and understandings of social interaction. Yet the academic discipline of communication has long suffered from a practical anthropocentrism that privileges human interaction and relegates the communication efforts of the more-than-human world to the margins of the discipline.[1] That many animals do indeed communicate—manipulating symbols, gesturing and even demonstrating a sense of self and other, has been argued at length by ethologists, zoologists, veterinarians, anthropologists, psychiatrists and biologists (e.g., Abram; Dawkins; Griffin; Mason; Midgley; Rogers and Kaplan; Sheldrake; Shepard; Zimmer). Gary Snyder puts it succinctly: "The evidence of anthropology is that countless men and women, through history and pre-history, have experienced a deep sense of communion and communication with nature and with specific non-human beings" (13).[2] As Jean Baudrillard points out, "animals were only demoted to the status of inhumanity as reason and humanism progressed" (29). Moreover, animals communicate in myriad ways that are, at least for most humans, either poorly understood or entirely unrecognized. Perhaps the gulf between some social and natural sciences and communication studies has contributed to the neglect of animal communication and human-animal communication, the subject of this book.

Our purpose in these chapters is to open up this area of investigation through consideration of a wide range of communication perspectives on human interactions with animals. We wish to do for communication studies what Chris Philo and Chris Wilbert did for geography in their insightful collection *Animal Spaces, Beastly Places*. More than this, though, we want to aid readers of all backgrounds in rethinking the role of communication in the construction and transformation of human relationships with the more-than-human world. Thus, the anthropocentric impulse holds fast in many of the chapters that follow, not to mention in some of our assumptions and understandings of animal communication. Bound in these pages by human language, that ancient art of rhetoric, we both recognize our limitations and hold them up for scrutiny. For example, some authors use the language of ownership to describe human relationships with companion animals while others make rhetorical choices that seek to challenge our ways of understanding interaction with other animals.[3] As Tema Milstein points out, "Struggles over discourse . . . are a necessary and interrelated part of wider struggles for change," including changes to human relationships with animals (1052). These chapters are but a starting point for consideration of the ways in which communication theories and methodologies can help us to broaden our critical horizons to include other species and, indeed, other worlds.[4]

Those approaching this volume with a foundation in the humanities and social sciences may recognize this call from the writings of several philosophers who have influenced the field of communication. Charles Saunders Peirce and George Kennedy, whose scholarship has been foundational for the study of rhetoric, offer invitations to consider animal communication. Their contributions are discussed briefly alongside an overview of extra-disciplinary scholarship that has also been influential in this area. Among the most notable semioticians to address the topic, for example, is Thomas Sebeok, whose various examinations of sign-based animal communication popularized the study of "zoosemiotics" or "biosemiotics" (Sebeok; Wheeler). Despite the 'human' bias in the communication field,[5] a few scholars have succeeded in publishing articles that explicitly address the subject of nonhuman communication (Barker; Carbaugh; Hawhee; Liska; Neiva and Hickson; Rogers; Rummel). Richard Rogers, in his germinal essay arguing for a materialist, transhuman and dialogic theory of communication, summarizes much of the relevant ecofeminist literature, highlighting the need for "ways of listening to nondominant voices and nonhuman agents and their inclusion in the production of meaning, policy, and material conditions" (268). As David Abram writes,

> To shut ourselves off from these other voices, to continue by our lifestyles to condemn these other sensibilities to the oblivion of extinction, is to rob our own senses of their integrity, and to rob our minds of their coherence. We are human only in contact, and conviviality, with what is not human. (Abram 22)

Critical theorists such as Gilles Deleuze and Felix Guattari provide substantial insight into the larger question of how animals and humans might communicate with each other, as do several ecofeminist authors (Adams; Gaard; Haraway; Merchant; Warren). Donna Haraway's *Companion*

Species Manifesto characterizes animals as "signifying others" (81); in it, she echoes anthropologist Barbara Noske's suggestion that we think about (and communicate with) animals as "other worlds" (34). Noske further suggests that ecofeminists, "unlike many other animal advocates . . . value non-animal nature, animate as well as inanimate" (Noske 173).[6]

For a semiotician such as Charles Saunders Peirce, feelings can "function as signs" (Anderson 86). He argued that animals have an instinct for communication and that the capacity to feel with another is the basis for perception. Clearly, certain animals signify with each other and across species, which Peirce described as "forms of communication . . . made possible by the shared feelings of difference perceivers" (qtd. in Anderson 87–88). Because of this ability to share feelings with others, Peirce suggests, like Kennedy, that we can "study the semeiotic, or sign-using, habits of all animals." (Anderson 87).

We are aided in doing so by expanding our understanding of communication beyond that very human obsession with the structure and substance of verbal utterances. Animals, including humans, speak not only via vocalization but also in scent, posture, eye gaze, even vibration. John Durham Peters describes communication as "the occasional touch of otherness" (256). For Kennedy, rhetoric is more than discursive; it is a "natural phenomenon: the potential for it exists in all life forms that can give signals, it is practiced in limited forms by nonhuman animals, and it contributed to the evolution of human speech and language from animal communication" (*Comparative Study* 4). Elsewhere, Kennedy argues that "rhetorical energy is not found only in language. It is present also in physical actions, facial expressions, gestures, and signs generally" ("A Hoot" 3–4).

Admitting that humans are generally inept at employing most systems of animal communication, Kennedy argues that we still "share a 'deep' natural rhetoric" with animals (*Comparative Study* 13). Through observation, we can "learn to understand animal rhetoric and many animals can understand some features of human rhetoric that they share with us, such as gestures or sounds that express anger or friendliness or commands" (*Comparative Study* 13). Kennedy's understanding of rhetoric suggests that communication is as much an exchange of energy as it is a matter of symbolic interaction (26). In fact, in his general definition of rhetoric, Kennedy alludes to the importance not only of acknowledging animal communication as rhetorical expression, but of enhancing the human interlocutor's ability to understand and take action.

> Rhetoric, in the most general sense, may thus be identified with the energy inherent in an utterance (or an artistic representation): the mental or emotional energy that impels the speaker to expression, the energy level coded in the message, and the energy received by the recipient who then uses mental energy in decoding and perhaps acting on the message. Rhetorical labor takes place. (*Comparative Study* 5)

On this last point, Barbara Noske points to several examples of humans who made an effort, who expended the rhetorical energy, to learn the language of their animal interlocutors and to listen to what they were expressing.[7] From a communication standpoint, such efforts demonstrate awareness of a point Noske makes shortly after addressing the question, "Is Animal Language not *Language?*"

> The basic question should not be whether animals have or have not human-like language. In having to pass *our* tests as measured by *our* yardsticks, they will always come out second best, namely, as reduced humans. The real question to be posed is how the animals themselves experience the world and how they organize this experience and communicate about it. (143–144).

Some of Noske's other arguments about human-animal communication are worth repeating here because, just as the subfield of intercultural communication has learned a great deal from anthropological studies of other humans, students of what I term *internatural communication* have much to gain from a critical anthropological approach to animal communication. Of particular note are Noske's observations regarding "feral" children raised by animals:

> In becoming one with the animals by virtually crossing the species boundary, these human beings not only have met the Other, they have almost become the Other. And by accepting this strange being in their midst the adoptive animals in their turn meet the Other. Indeed, animal-adopted children exemplify an animal-human relationship more than a human-animal relationship. . . .
>
> Even though we may not succeed in becoming animal with the animals, we as humans may make the effort of meeting the animals on their own ground instead of expecting them to take steps towards us and making them perform according to our standards. . . . To do this one must try to empathize with animals, to imagine what it is to be a wolf, a dolphin, a horse or an ape. (167)

She goes on to say, "Good participatory observation is basically an exercise in *empathy* while at the same time one is aware of the impossibility of total knowledge and total understanding" (169). It is this empathic impulse that drives this collection.[8]

Deleuze and Guattari's essay "Becoming-Animal," published in *A Thousand Plateaus*, provides a way of thinking about communication that, in some ways, echoes Noske's call for empathy and Kennedy's definition of rhetoric as essentially "a form of mental and emotional energy" (*Comparative Study* 3). For Deleuze and Guattari, "becoming-animal" is about movement and proximities. "Becoming is to emit particles that take on certain relations of movement and rest because they enter a particular zone of proximity. Or, it is to emit particles that enter that zone because they take on those relations" (122). They offer instructions for grasping this notion of human-animal compossibility, this "shared and indiscernible" proximity "that makes it impossible to say where the boundary between the human and the animal lies" (122).

> An example: Do not imitate a dog, but make your organism enter into composition with *something else* in such a way that the particles emitted from the aggregate thus composed will be canine as a function of the relation of movement and rest, or of molecular proximity, into which they enter. Clearly, this something else can be quite varied, and be more or less directly related to the animal in question . . . (Deleuze and Guattari 123)

Later in the essay, the authors affirm the molecular nature of "becoming-animal": "Yes, all becomings are molecular: the animal, flower or stone one becomes are molecular collectivities, haecceities, not molar subjects, objects, or forms that we know from the outside and recognize from experience, through science, or by habit" (124).

More than mimicry or reflection, though, this 'becoming' is a manifestation of corporeal dialogism, an "embodied rhetoricity" and perspective on communication that "forsakes oppositionality in favor of an all-encompassing perspective on the rhetorical act" (McKerrow 319). Emphasizing corporeality, Deleuze and Guattari suggest we must allow ourselves to feel, at a molecular level, the connection to otherness. In the process of becoming-molecular, becoming-animal, we humans might do well to attend to other sources of meaning and intentionality with the same scrutiny and care we give to the symbolic. As animal behaviorist V. Csanyi points out,

> The "top-down" approach, which compares animal accomplishments to those of humans, is heavily burdened with ideology. . . . Of course, there are other avenues as well. We could examine, for example, how animals, or even humans, *understand* how one should behave in a small community. . . . A true evolutionary characterization would adopt such an approach. (Csanyi 167)

Drawing insight and encouragement from these and other theorists who see no reason *not* to consider communication as, at the very least, an interspecies enterprise,[9] I offer this collection as a foray into the realm of *internatural communication*. It is a first step toward what I expect will grow into a more expansive set of questions about communication and the more-than-human world.[10] Like intercultural communication's emphasis on relationships among and between different cultures, internatural communication explores interaction among and between natural communities and social groups that include participants from what we might initially describe as different classifications of nature.[11] Internatural communication includes the exchange of intentional energy between humans and other animals as well as communication among animals and other forms of life. It is at its core, as is the study of communication generally, about the construction of meaning and the constitution of our world through interaction. It simply extends the boundary line a little further, first to include other animals, so that we can test the veracity and capacity of our theories and methods in this new space.

Similar to Kennedy's approach to the 'rhetorical study of animal communication,' which focuses on the identification of principles and formal aspects of communication commonly used by both human and nonhuman animals, the authors in this volume approach animal-human communication questions from standpoints shaped by communication theories and research methods. Some of this work might elsewhere be termed "zoosemiotics," "biorhetoric," "communibiology," "ecosemiotics," "anthrozoology"[12] or even "corporeal rhetoric" or "transhuman communication." I choose the term *internatural communication* not to compete with these other labels but rather as a term that can be inclusive of their meanings as well as embracing the possibilities of human and animal communication with other life forms. I also like the term because of its capacity to capture a way of communicating with and about nature from a standpoint that is implicated in the very concept of 'nature.'

The organization of this book reflects a perspective on communication informed by a coherentist epistemology. Such an approach "privileges no one position at the expense of others because it begins with the assumption that all positions are interrelated and interdependent" (McPhail, "From Complicity to Coherence" 127). According to Mark Lawrence McPhail, inquiry into coherence begins with "a radical critique of duality" and moves toward an "emancipatory understanding of language and life" (*Zen* 5–6). In addition, it emphasizes the kind of "methodological and epistemic flexibility" characteristic of this volume ("From Complicity to Coherence" 127). We begin, then, by examining the question of our complicity in the rhetorical, ideological and practical subordination of animals and animal subjectivity to human interests and agendas.[13] From there, we move along a continuum of essays in Part II that ask us to consider our implication in the lives of animal Others. McPhail describes implication as "the recognition and awareness of our essential interrelatedness . . . " (*Rhetoric* ix). Some of the essays in this section are aimed at extending communication theory to address the significance of human-animal relationships for the humans (and sometimes other animals) in those relationships. Other chapters focus on implication as the praxis for coherence, a process for coming to relate, listen and interact in ways that honor the integrity of animals and our relationships with them. Finally, in Part III, we explore the possibilities of a coherence theory of human-animal relations through explorations of internatural communication in both domestic and wild contexts, as well as through arguments for repositioning our human ways of communicating and knowing alongside, rather than above, those of other animals. The book concludes by calling for efforts to expand our understanding of internatural communication by rethinking our anthropocentric grip on the symbolic and becoming students of corporeal rhetorics of scent, sound, sight, touch, proximity, position and so much more.

Following the introduction, Part I opens with a theoretical essay by Tony Adams in which he argues that human representations of animals mediate "personal and political human agendas" in ways that naturalize those agendas. Grounding his analysis in symbolic interactionism, Adams weaves together personal experience, textual analysis and ethnographic fieldwork to show how humans use companion animals to mediate interaction with other humans, how the Central Park Zoo's purportedly gay penguin couple (and other popular penguin depictions) mediates public discourse on gay marriage and how invasive species displays at aquariums and zoos mediate human dialogues about immigration policy. In Chapter 3, Deborah Cox Callister examines the rhetoric surrounding beached whales in order to understand how the bodies and circumstances of the whales shape and influence human understanding and orientation toward particular policy objectives. At a time when the U.S. Navy proposes five years of testing and training of sonar and explosives that threaten millions of marine mammals in the Atlantic and Pacific Oceans, Callister's analysis and materialist rhetorical perspective is especially significant. Elaborating the policy perspective, Joseph Abisaid takes on the primate research debate in Chapter 4. Abisaid conducts a framing analysis of the primate testing debate, arguing that appeals to scientific progress continue to counter the ethical arguments against experimentation on nonhuman primates, providing insight into "how individuals rationalize the human-animal relationship." Moving away from questions of policy and toward the ideological orientations that undergird such decision making, Shana Heinricy's chapter, which concludes the section, examines animal representation as configured in the history of American animation. Using a case study of the popular children's program *SpongeBob Squarepants*, Heinricy argues that visual representations of animated animal bodies often "create, maintain, and render invisible speciesist ideologies."

The essays in Part II move us from complicity to implication, a critical awareness and effort to understand and make our role as humans in communicative relationships and interactions with other animals more just and responsible. In Chapter 6, Carrie Packwood Freeman calls attention to the necessary connections between human values and food consumption in arguing for a vegan ethic. Like Wendy Atkins-Sayre's recent essay examining how PETA seeks to overcome the human-animal divide, Freeman's chapter notes that the rhetoric of prominent animal rights organizations works to overcome human views of animals as Others, especially animals used for food. She illustrates how several groups advocate for veganism by appealing to primary human values and evaluates the campaigns in terms of how effectively they challenge speciesism and accomplish their goal of persuading consumers. In fact, the incongruity in humans' treatment of animals viewed as a source of food and animals viewed as companions is significant. Thus, our effort to become further implicated in questions of communication in human-animal relationships turns next to those animals known commonly as 'pets.'

Nick Trujillo takes his companion Ebbie on the road in Chapter 7 to learn about "dog culture" and the culture of "dog people." His essay illustrates the power of the canine-human bond and provides insight into the variety of human communication practices related to living with and serving those with such bonds. In Chapter 8, Mary Pilgram looks at supportive communication from the veterinarian's point of view, driving home Trujillo's point that many people consider their animals to be (and to be treated as) members of their families. Pilgram investigates veterinarians' perceptions of their social support efforts toward grieving human clients, suggesting ways that training in supportive communication could enhance professional practice.

Turning from companion animals to wildlife, the last chapters offer a unique perspective on humans' implication in environmental (un)sustainability. Leigh Bernacchi looks at the ritual interaction that unites birders and birds and suggests ways the relationship can be extended toward a conservation ethic in Chapter 9. Her argument is reminiscent of Kennedy's statement about bird songs: "Ritualization accompanied by epideictic utterance is a feature of animal rhetoric as it is of human life" (21). Concluding the section on implication and pointing the way toward coherentist perspectives, Tema Milstein critiques the contemporary "naturalistic" zoo in Chapter 10. She examines the zoo in terms of its institutionalized practices of reflection. She then explores possibilities for rhetorical refraction, introduced by the young visitors and inhabitants of the zoo exhibits who challenge the human-animal divide.

In Part III, our coherentist framework comes full circle, illustrating how communication studies can move beyond a focus solely on interactions among humans to be an interspecies and international enterprise, as is the world for which it seeks to account. The first chapter in the section uses ethnographic methods to study the techniques and outcomes of animal-human communicators, professionals who communicate (and train other humans to communicate) with particular animals. Even skeptics of animal-human communicators are likely to find Susan Hafen's evidence and argument for more affective and intuitive communicative processes compelling. In Chapter 12, Pat Munday looks at ravens and human hunters from the perspective of semiotics, arguing that ravens' relationships with other animals (including humans) as well as their communicative capacities can help us understand how to bridge the animal-human divide. Further legitimating

Munday's arguments, Stephen Lind makes the case for symbolic animal communication in Chapter 13, challenging Kenneth Burke's definition of humans as the only symbol-using animal and opening up consideration of the "complicated and fascinating ways" animal-human communication can function. The book concludes with Susannah Bunny LeBaron's rumination on the narrative elements of human communication with the more-than-human world, in which she argues for a paradigm shift in the ways we narrate the constitutive relationship between humans and other animals. Her 'meditation' reminds us that implication is a path toward coherence, both narrative and ideological, in our interactions with other animals.

Whether in the lab, the field, a zoo, at home, at the vet, in the garden, on the road or on the screen, we regularly encounter communicating animals and often direct our mental energies toward them. As these essays demonstrate, communication researchers should have a lot to say about the dynamics of human-animal communication. Even more importantly, we are well poised to offer new (and old) ways of listening and learning, internaturally:

> / Remember the plants, trees, animal life who all have their / tribes, their families, their histories, too. Talk to them, / listen to them. They are alive poems /

> (Joy Harjo 40)

NOTES

1. Stephen Lind elaborates on this observation in Chapter 13, this volume. Like other authors in this volume, I borrow the phrase "more-than-human world" from David Abram, who uses it in reference to "sensuous reality" (x).
2. For a popular description of some of these communicative relationships, see Deborah Noyes's *One Kingdom*.
3. Consider, for instance, Donna Haraway's challenge to linguistic (and ideological) dualisms with terms such as "humanimal" and "natureculture."
4. Barbara Noske's *Beyond Boundaries* is a germinal volume for any student of human-animal communication. She points out how the "bias of human domination" contributes to "an ideological stake in a status quo: the object status of animals" (101). In contrast, she views culture as a "dialectical process of constituting and being constituted," a process that frequently involves relationships across boundary lines termed *species* (87). Looking at similar issues from a representational standpoint, Stacey Sowards draws upon Kenneth Burke's concept of consubstantiality to explain how identification with orangutans can "deconstruct the nature/culture divide and dualistic thinking that has persisted for centuries" (46). She argues that "animalcentric anthropomorphism" can provide a "profound interspecies event" that is inclusive of human-animal continuities as well as discontinuities (46).
5. Even Celeste Condit has acknowledged that rhetorical critics must move beyond our "ethnocentric assumption that only human-made symbolic codes matter to human action" (371).
6. Although this volume focuses exclusively on the human-animal relationship, nonanimal nature can also be understood within the framework of

internatural communication I propose. McKerrow gestures toward this possibility when he extends his concept of corporeal rhetoric to explain Randall Lake's description of "Red Power" rhetoric, which holds land as an "essential element of Indian identity" (324). By transcending the nature/culture dualism, corporeal rhetoric can enable an expanded notion of relationship and recognition; in short, corporeal rhetoric may serve as a powerful resource for internatural communication.

7. See Donal Carbaugh's essay "Just Listen" for a communication studies example of such deep listening.

8. Karen Dace and Mark McPhail, writing about interracial interaction, offer empathy as a communicative behavior that can lead to "implicature," or "the notion that human beings are linguistically, materially, psychologically, and spiritually interrelated and interdependent" (345–346). I argue that such implicature (or "implication," as I refer to it in this volume) can be practiced with the more-than-human world as well.

9. Noske also makes the case for interspecies communication (156). Carl Zimmer's recent *Time* cover story on "Animal Friendships" mentions "one of the most provocative implications" of recent research into animal friendships, namely "that friendships that evolved within species may sometimes reach across the species barrier" (38).

10. Actually, it is perhaps better characterized as a second or third step, as other communication scholars have helped to establish this trail (see, e.g., Michael Salvador and Tracylee Clarke's essay on "The Weyekin Principle" and Julie Schutten and Richard Rogers's essay on "transhuman dialog" [*sic*]).

11. Of course, the very notion of classification breaks down under further scrutiny. What lines of difference will matter at any given historical moment? Mammal or animal? Vertebrate or invertebrate? Flora or fauna? Skin or shell or fur? Animate or inanimate?

12. For a more detailed discussion of anthrozoology, see Susan Hafen's chapter (this volume).

13. For an extensive discussion of the sociological dimensions and development of human relationships with and use of animals in the twentieth century, see Adrian Franklin's *Animals and Modern Cultures*.

REFERENCES

Abram, David. *The Spell of the Sensuous: Perception and Language in a More-Than-Human World*. New York: Pantheon Books, 1996. Print.

Adams, Carol J. *Neither Man nor Beast: Feminism and the Defense of Animals*. New York: Continuum, 1995. Print.

———. *The Sexual Politics of Meat: A Feminist-Vegetarian Critical Theory*. New York: Continuum, 1990. Print.

Anderson, Douglas R. "Peirce's Horse: A Sympathetic and Semeiotic Bond." *Animal Pragmatism: Rethinking Human-Nonhuman Relationships*. Ed. Erin McKenna and Andrew Light. Bloomington: Indiana UP, 2004. 86–94. Print.

Atkins-Sayre, Wendy. "Articulating Identity: People for the Ethical Treatment of Animals and the Animal/Human Divide." *Western Journal of Communication* 73.2 (2010): 309–328. Print.

Barker, Randolph T. "On the Edge or Not? Opportunities for Interdisciplinary Scholars in Business Communication to Focus on the Individual and Organizational Benefits of Companion Animals in the Workplace." *Journal of Business*

Communication 42.3 (2005): 299–315. *Ebscohost Communication and Mass Media Complete.* Web. 4 Apr. 2010.

Baudrillard, Jean. "The Animals: Territory and Metamorphoses." *Simulacra and Simulation.* Trans. Sheila Faria Glaser. Ann Arbor: U of Michigan P, 1994. Print.

Bekoff, Marc. *Minding Animals: Awareness, Emotion, and Heart.* Oxford: Oxford UP, 2002. Print.

———. *The Emotional Lives of Animals.* Novato: New World, 2007. Print.

———. *The Animal Manifesto: Six Reasons for Expanding Our Compassion Footprint.* Novato: New World Library, 2010. Print.

Carbaugh, Donal. "'Just Listen': 'Listening' and the Landscape among the Blackfeet." *Western Journal of Communication* 63.3 (1999): 250–270. Print.

———. "Quoting the Environment: Touchstones on Earth." *Environmental Communication: A Journal of Nature and Culture* 1.1 (2007): 64–73. Print.

Condit, Celeste. "Contemporary Rhetorical Criticism: Diverse Bodies Learning New Languages." *Rhetoric Review* 25.4 (2006): 368–372. Print.

Csányi, Vilmos. *If Dogs Could Talk: Exploring the Canine Mind.* Trans. Richard E. Quandt. New York: North Point P, 2005. Print.

Dace, Karen L., and Mark L. McPhail. "Crossing the Color Line: From Empathy to Implicature in Intercultural Communication." *Readings in Intercultural Communication: Experiences and Contexts.* 2nd ed. Ed. Judith N. Martin, Tom K. Nakayama, and Lisa A. Flores. Boston: McGraw Hill, 2002. 344–350. Print.

Dawkins, Marian Stamp. *Through Our Eyes Only?* New York: Oxford UP, 1998. Print.

Deleuze, Gilles, and Félix Guattari. *A Thousand Plateaus.* Trans. Brian Massumi. Minneapolis: U of Minnesota P, 1987. Print.

Derrida, Jacques. *The Animal That Therefore I Am.* Ed. Marie-Louise Mallet. Trans. David Wills. New York: Fordham UP, 2008. Print.

Franklin, Adrian. *Animals and Modern Cultures: A Sociology of Human-Animal Relations in Modernity.* Thousand Oaks, CA: Sage, 1999. Print.

Gaard, Greta, ed. *Ecofeminism: Women, Animals, and Nature.* Philadelphia: Temple UP, 1993. Print.

Grandin, Temple. *Animals in Translation: Using the Mysteries of Autism to Decode Animal Behavior.* New York: Scribner, 2005. Print.

Griffin, Donald R. *Animal Thinking.* Cambridge: Harvard UP, 1984. Print.

———. *Animal Minds: Beyond Cognition to Consciousness.* Chicago: U of Chicago P, 2001. Print.

Haraway, Donna. *Primate Visions: Gender, Race, and Nature in the World of Modern Science.* New York: Routledge, 1989. Print.

———. *The Companion Species Manifesto: Dogs, People, and Significant Otherness.* Chicago: Prickly Paradigm P, 2003. Print.

———. *When Species Meet.* Minneapolis: U of Minnesota P, 2008. Print.

Harjo, Joy. "Remember." *She Had Some Horses.* New York: Thunder's Mouth P, 1983. Print.

Hawhee, Debra. "Toward a Bestial Rhetoric." *Philosophy & Rhetoric* 44.1 (2011): 81–87. Print.

Irigaray, Luce. "Animal Compassion." Trans. Marilyn Gaddis Rose. *Animal Philosophy: Ethics and Identity.* Ed. Matthew Calarco and Peter Atterton. London: Continuum, 2004. 195–201. Print.

Kennedy, George A. "A Hoot in the Dark: The Evolution of General Rhetoric." *Philosophy and Rhetoric* 25.1 (1992): 1–21. Print.

———. *Comparative Rhetoric: An Historical and Cross-Cultural Introduction.* New York and Oxford: Oxford UP, 1998. Print.

Kull, Kalevi. "A Note on Biorhetorics." *Sign System Studies* 29.2 (2001): 693–704. Print.

Liska, Jo. "The Role of Rhetoric in Semiogenesis: A Response to Professor Kennedy." *Philosophy and Rhetoric* 26.1 (1993): 31–38. Print.

Mason, William A. "Development of Communication between Young Rhesus Monkeys." *Science* 130 (1959): 712–713. Print.

McKenna, Erin, and Andrew Light, eds. *Animal Pragmatism: Rethinking Human-Nonhuman Relationships*. Bloomington: Indiana UP, 2004. Print.

McKerrow, Raymie E. "Corporeality and Cultural Rhetoric: A Site for Rhetoric's Future." *Southern Communication Journal* 63 (1998): 315–328. Print.

McPhail, Mark Lawrence. "Complicity: The Theory of Negative Difference." *The Howard Journal of Communication* 3 (1991): 1–13. Print.

———. *Zen in the Art of Rhetoric: An Inquiry into Coherence*. New York: State U of New York P, 1996. Print.

———. "From Complicity to Coherence: Rereading the Rhetoric of Afrocentricity." *Western Journal of Communication* 62 (1998): 114–140. Print.

———. *Rhetoric of Racism Revisited: Reparations or Separation?* Lanham: Rowman & Littlefield, 2002. Print.

McReynolds, Phillip. "Overlapping Horizons of Meaning: A Deweyan Approach to the Moral Standing of Nonhuman Animals." *Animal Pragmatism: Rethinking Human-Nonhuman Relationships*. Ed. Erin McKenna and Andrew Light. Bloomington, IN: Indiana UP, 2004. 63–85. Print.

Merchant, Carolyn. *The Death of Nature: Women, Ecology and the Scientific Revolution*. San Francisco, CA: Harper & Row, 1980. Print.

Midgley, Mary. *Animals and Why They Matter*. Athens, GA: Georgia UP, 1983. Print.

Milstein, Tema. "Human Communication's Effect on Relationships with Animals." *Encyclopedia of Human-Animal Relationships: A Global Exploration of Our Connections with Animals*. Ed. Marc Bekoff. Vol. 3. Westport: Greenwood P, 2011. 1044–1054. Print.

———. "When Whales 'Speak for Themselves': Communication as a Mediating Force in Wildlife Tourism." *Environmental Communication: A Journal of Nature and Culture* 2.2 (2008): 173–192. Print.

Neiva, Eduardo, and Mark Hickson III. "Deception and Honesty in Animal and Human Communication: A New Look at Communicative Interaction." *Journal of Intercultural Communication Research* 32.1 (2003): 23–45. *Ebscohost Communication and Mass Media Complete*. Web. 4 Apr. 2010.

Noske, Barbara. *Beyond Boundaries: Humans and Animals*. Montreal: Black Rose Books, 1997.

Noyes, Deborah. *One Kingdom: Our Lives with Animals: The Human-Animal Bond in Myth, History, Science, and Story*. Boston: Houghton Mifflin, 2006. Print.

Peters, John Durham. *Speaking into the Air: A History of the Idea of Communication*. Chicago: U of Chicago P, 1999. Print.

Philo, Chris, and Chris Wilbert, eds. *Animal Spaces, Beastly Places: New Geographies of Human-Animal Relations*. London and New York: Routledge, 2000. Print.

Plumwood, Val. *Feminism and the Mastery of Nature*. New York: Routledge, 1993. Print.

Rogers, Lesley J., and Gisela Kaplan. *Songs, Roars, and Rituals: Communication in Birds, Mammals, and Other Animals*. Cambridge: Harvard UP, 2000. Print.

Rogers, Richard. "Overcoming the Objectification of Nature in Constitutive Theories: Toward a Transhuman, Materialist Theory of Communication." *Western*

Journal of Communication 62 (1998): 244–272. *Ebscohost Communication and Mass Media Complete.* Web. 4 Apr. 2010.

Rummel, Lynda. "Animal and Human Communication." *Eastern Communication Association Conference*, New York City: 14 Mar. 1975. N. pag. *Ebscohost Communication and Mass Media Complete.* Web. 4 Apr. 2010.

Salvador, Michael, and Tracylee Clarke. "The Weyekin Principle: Toward an Embodied Critical Rhetoric." *Environmental Communication: A Journal of Nature and Culture* 5.3 (2011): 243–260. Print.

Schutten, Julie, and Richard Rogers. "Magick as an Alternative Symbolic: Enacting Transhuman Dialogs." *Environmental Communication: A Journal of Nature and Culture* 5.3 (2011): 261–280. Print.

Sebeok, Thomas A. "Zoosemiotics." *American Speech* 43.2 (1968): 142–144. *Ebscohost Communication and Mass Media Complete.* Web. 4 Apr. 2010.

———. *Global Semiotics.* Bloomington: Indiana UP, 2001. Print.

Serpell, James. *In the Company of Animals: A Study of Human-Animal Relationships.* Cambridge: Cambridge UP, 1986. Print.

Sheldrake, Rupert. *Dogs That Know When Their Owners Are Coming Home.* New York: Crown Publishing Group, 2011. Print

Shepard, Paul. *Thinking Animals: Animals and the Development of Human Intelligence.* Athens: U of Georgia P, 1998. Print.

Singer, Peter. *Animal Liberation.* 2nd ed. New York: Random House, 1990. Print.

Snyder, Gary. *The Old Ways: Six Essays.* San Francisco: City Lights Books, 1977. Print.

Sowards, Stacey. "Identification through Orangutans: Destabilizing the Nature/Culture Dualism" *Ethics & the Environment* 11.2 (2006): 45–61. Print.

Warren, Karen. *Ecofeminist Philosophy: A Western Perspective on What It Is and Why It Matters.* Lanham: Rowman and Littlefield, 2000. Print.

Wheeler, Wendy. *The Whole Creature: Complexity, Biosemiotics and the Evolution of Culture.* London: Lawrence and Wishart, 2006. Print.

Zimmer, Carl. "The Surprising Science of Animal Friendships." *Time* 20 Feb. 2012: 34–38. Print.

Part I
Complicity

2 Animals as Media
Speaking through/with Nonhuman Beings

Tony E. Adams

Two days before Halloween and four days before the 2004 U.S. presidential election, I sit outside my favorite coffee shop and prepare to read the newspaper. I am quickly distracted by people walking decorated dogs, animals sporting costumes and political propaganda. I watch a Doberman in a miniature cowboy hat and a pumpkin-suited Chihuahua pass, followed by a Labrador with a John Kerry/John Edwards bumper sticker attached to its fur.

I notice that the decorated dogs and their owners receive more attention than owners of undecorated dogs; the decorated dogs seem to work as conversation starters, separating the animals and their owners from the mundane and boring owners and animals. The Labrador in support of Kerry/Edwards also functions as a vehicle to take the owner's political message into narrow alleys, sidewalks and dog-friendly venues.

In this essay, I describe how animals can function as media, as tools humans use to facilitate human interaction. In so doing, I add to existing research on human-animal relationships, research that tends to emphasize the dilemmas that arise when humans treat animals as people, objects or a combination of both (Francione; Sanders); ways humans speak for animals, ways animals communicate with people and ways humans can and should interpret animal communication (Arluke and Sanders); what animal behaviors tell us about human behaviors (MooAllem; Roughgarden); ways (human) representations of animals can influence human interactions with and communication about live versions of these animals (Berger; King); animal selfhood and the "shared intersubjectivity" of humans and animals (Jerolmack 655; Irvine); and the mutual, coevolving qualities of "companion species" relationships, meaningful endeavors created by all species involved (Anderson; Haraway, "Species").

Some writers have acknowledged ways humans use animals as media. For instance, Cain describes how humans talk "to their pet instead of to other family members" in ways "other family members could hear" (79); Messent refers to dogs as "social lubricants" (45); Williams frames dogs as "relational media" (103); Arluke and Sanders provide examples of people presenting a "virtual voice" of an animal to express their own "orientation, desires, or concerns" (70); Robins, Sanders and Cahill demonstrate ways an animal can work as a "conduit" humans use to speak to other humans (22); and Ramirez illustrates how humans can use dogs as "props" in order to create "presentations of self" (375). However, the specific ways humans use animals as media and the implications of such use are tangential discussions in many of these projects.

Given my interest in understanding ways humans use animals as media in human interaction, tenets of symbolic interactionism ground this research. Interactionists concern themselves with what happens in moments of relating, in the time and space of interaction. In particular, interactionists work to discern the "taken-for-granted meanings" entrenched in interaction processes (Denzin 19), attend to meaning-making processes (Goffman, "Interaction"; Mead), conceive of personal accountability in interaction (Goffman, "Strategic"; Scott and Lyman) and demonstrate how meanings are used, by humans, to make sense of themselves, others and society.

Adhering to interactionist goals, I have two interrelated objectives. First, I describe what it means to make animals media, and specifically note how humans, *in* interaction, make animals meaningful *for other humans*. I use two case studies to formulate this animals-as-media theory: (1) the use of dogs by humans and (2) the use of penguins at the Central Park Zoo (New York) and in the film *March of the Penguins*. I then discern possible consequences of using animals as media by illustrating how the rhetoric of "invasive species" exhibits found at many zoos and aquaria can implicate humans labeled "illegal," "invasive" and "nonnative."

Second, following Irvine's call for researchers to better understand "how" animals mean something for human interaction (15), I illustrate how the human use of animals can influence meaning-making processes and personal accountability. As I demonstrate, animals are not "neutral delivery system[s]," an assumption often held about media (Meyrowitz 15). Animals can, and do, harbor personal and political human agendas.

METHOD

A case study is a detailed account of an activity or a process. The purpose of the account is to provide insight into, advance theorizing about and attend to the social and political characteristics of the activity or process (Stake). Case studies utilize multiple methodological procedures and sources of evidence (Yin), and they are helpful because they refine theory and introduce complexities for future research (Creswell).

In this project, I use two case studies—the use of dogs by humans, and the use of gay and straight penguins—to provide an account of how humans can use animals as media. I use personal experience, textual analysis and ethnographic fieldwork to develop each case. I then use grounded theory (Charmaz) to inductively discern patterns—repeated words, phrases and topics of discussion—across my two cases in order to theorize human animal-use.

As my opening narrative suggests, this project emerged from observations of people using dogs as interactional devices in everyday settings. I treated these observations as meaningful and valuable, and decided to research human dog-use. My first case thus describes ways humans can make dogs meaningful for other humans.

Using phrases like "dog clothes," "dressed up animals" and "dogs and human interaction," I searched online newspaper, magazine and journal databases to find stories of humans communicating with other humans through the use of dogs. I wanted examples that complemented my personal experience and initial observations. I was not interested in articles about human interpretations of dog communication (or dog interpretations of human communication), whether dogs and humans can meaningfully relate or the ethics of humans speaking for dogs.

While I was working on human dog-use, controversy emerged around Roy and Silo, a same-sex penguin couple at the Central Park Zoo (New York), their accompanying children's book *And Tango Makes Three* and penguins featured in the film *March of the Penguins*. In this controversy, I noticed a similarity between the human use of penguins and the human use of dogs reported in my opening anecdote. I thus decided to research articles about human penguin-use as well.

Using phrases like "gay penguins," "Roy and Silo," "*And Tango Makes Three*" and "*March of the Penguins*," I searched online newspaper and magazine databases to find stories of human penguin-use. Similar to my dog search, I wanted articles that discussed ways humans made penguins meaningful for other humans. I was not interested in articles that that announced the release of the book or the film or articles documenting how well the book or film sold.

While doing this research, I was also conducting ethnographic fieldwork at the Florida Aquarium (Tampa). I soon noticed that the rhetoric of the invasive species exhibit resembled ways humans were using dogs and penguins. But it seemed that the rhetoric of these exhibits was much more politicized, particularly because it called for the eradication of animals labeled "illegals," "invaders" and "nonnatives," labels that, when viewing animals as media, could have consequences for humans labeled in similar ways. Therefore, I found it important to illustrate possible consequences of such rhetoric by discussing not only the discourse at the Florida Aquarium, but also the discourse at other invasive exhibits around the United States.

MAKING ANIMALS MEDIA

Media are "agencies" that enable "communication to take place" (Fiske 176–177). Media are vehicles humans use to communicate goals and pleasures, vehicles that harbor values and ideologies (Lorde; Spigel). Media are sensory "extensions" of human bodies, extensions that make new and unique ways of human interaction possible (McLuhan; Ong; Schivelbusch). And media facilitate human "sense-making" (Pauly).

In this essay, I frame animals as media, as agencies humans can use to speak to other humans and agencies that enable human interaction. Animals function as media when humans use them to convey information to other humans, as the Labrador sporting the Kerry/Edwards bumper sticker demonstrates. When clothed, decorated animals signify, for other humans, creative and responsible owners (Prager). Conceiving of animals as media provides another way to make sense of human-animal relationships and, more importantly, makes animals important interactional phenomena.

This animals-as-media theory may initially resemble anthropomorphism—"the projection onto animals of our own feelings, thoughts, motives, and other human qualities" (Challinor 1). However, the theory differs from anthropomorphism in that an animal need not be humanized to function as a medium. Humans can make "statements" *with* an animal *for* other humans, and such statements may not involve ascribing human qualities to an animal. In some instances, anthropomorphism may occur, but it is not necessary for animals to function as media.

Next, I describe how dogs can "speak for" their owners. As I demonstrate, these ways of speaking are made possible only by the presence of dogs and are ways of speaking that are meaningful for and directed at humans, not necessarily for or at other dogs. I also illustrate how people can evaluate a person based on the type of animal the person owns and how the person cares for a particular animal.

MAKING ANIMALS MEDIA

Prager asks, "Do owners dress their dogs because it's cold outside? Or just because it's cool?"(V1). She writes of dressing nonhuman beings and considers animal attire socially important: "leaving the house spells clothing to human beings. It's a lesson learned from infancy: you dress to go out. You have a furry creature with you, you dress it, too" (V8; see Levine). Prager suggests that humans dress animals because it is socially appropriate—that is, appropriate for humans. Based on this logic, naked animals can signify inappropriate and irresponsible owners. It is this signifying possibility that makes animals media.

Or consider the following statements made in a human argument: "'Stop fighting—you're upsetting the dog,' or 'Get quiet—the dog is upset'" (Cain 76). In an attempt to get someone to stop fighting or to get quiet, a person uses a dog to justify a request. The argument may have nothing to do with the dog itself, but the person's request and justification of the request is based on the presence and anthropomorphizing of a dog. But here the dog functions as a medium, an agency that enables a particular discourse to take place.

Or consider a question posed by Tapper: "Can we judge a man by his dog?" (12). Tapper describes how the types of dogs U.S. presidents own and the presidents' treatments of these dogs suggest to other humans—and not to other dogs—what the presidents are like as humans. Here dogs, by being owned and acted upon by humans, speak for humans to humans. Tapper specifically focuses on Mitt Romney's transporting of Seamus, the family dog, in a cage on top of the car. Based on Romney's inhumane treatment, people wondered if Romney was compassionate. Tapper also illustrates how a similar evaluation occurred when Lyndon Johnson lifted his beagles by their ears, when it was revealed that Bill Frist adopted cats from humane shelters in order to conduct experiments on them and when Barack Obama broke a campaign pledge of adopting a rescue mutt by instead adopting a pure breed (Tapper and Miller). In these examples, anthropomorphism did not apply: People did not give dogs and cats *human* qualities. Rather, dogs and cats—and human acts performed in relation to these dogs and cats—mediated human evaluations of humans.

Or consider Ramirez's observation that the presence of a dog next to a person is often indicative, for others, of this person's masculinity (cats, Ramirez suggests, are indicative of femininity). Miller adds another layer to Ramirez's claim. He notes that because of the "froufrou" factor associated with poodles, few men "want to be seen parading down the street with one" (16). Miller suggests that, by adhering to gendered stereotypes, some men believe others will find them less masculine if they walk a prettified dog, less masculine if they attach a poodle to a leash and walk it in public settings. Here, concern does not reside with the dogs themselves but with the owners and (perceived) human evaluations of the owners. Ramirez and Miller demonstrate that by (not) using dogs, some men articulate and perpetuate gender "statements" by showing others (and themselves) that they can "do" masculinity well, that is, by walking a dog (and not a cat) or by refusing to walk a "froufrou" dog.

Human attributions of meaning can also affect the human-animal relationship. For instance, if a man considers a cat feminine or a poodle unmanly and thus does not want to be seen with either emasculating being, the cat and the poodle could suffer from the man's neglect. Anthropomorphism does not apply in these examples either: This man did not give the cat or poodle *human* qualities. The man invested a cat or poodle with meanings important for himself and possibly other humans; the animal only functioned as a medium, as a vehicle that transmitted particular gendered information.

While I have illustrated how particular dogs and cats might function as media, I next introduce a more politicized example of human animal-use. I specifically show how some people used straight and gay penguins to justify (human) arguments and political causes, arguments and causes that, while made possible by the presence of penguins, have no direct bearing on the everyday existence of the penguins.

PENGUINS AS MEDIA

June 10, 2002. Freelance journalist Cristina Cardoze introduces Wendell and Cass, a pair of male penguins living at the New York Aquarium. At this time, Wendell and Cass have been coupled for eight years. The penguins' caretaker, Stephanie Miller, says, "There are a lot of animals that have same-sex relations, it's just that people don't know about it." This lack of knowledge makes Wendell and Cass newsworthy. "At the New York Aquarium, no one suspected Wendell and Cass were gay when they first bonded," Cardoze explains. "Penguins don't have external sex organs, so visually there's no surefire way to tell whether they are male or female." The birds regularly copulate but never produce an egg, and Miller, the caretaker, debunks the myth that Wendell and Cass, because they are gay, have the cleanest nest: "These are penguins. They poop in their nest. Nobody's got a clean nest."

Humans categorize animals on human terms. Animals are deemed "ugly" or "cute," "clean" or "dirty," "female" or "male." When it comes to sexuality, however, animals are often categorized as "heterosexual." Based upon human definitions of sexuality, nonheterosexual animals have always existed, but, because of conservative politics and "histories of censorship" (Adam 18), they have only recently appeared in scientific studies (see Bagemihl; Chris; Giffney and Hird; MooAllem; Reiss; Terry), in children's books (Richardson and Parnell) and on television (Alexandresco, Loyer, and Menendez; Minhas and Littleboy; Thomas). Nonhetero species still remain absent from most museums, zoos and aquaria (Desmond; Liddiard) and many nature documentaries exclude them as well (Uddin and Hobbes). Enter Roy and Silo.

February 7, 2004. Dinitia Smith, a reporter for the *New York Times*, tells of a "love that dare not squeak its name." It is the story of Roy and Silo, a male penguin couple residing at the Central Park Zoo in New York. Smith notes that the birds have been partnered for six years, and, out of a desire to incubate an egg, "put a rock in their nest and sat on it, keeping it warm in the folds of their abdomens." Rob Gramzay, the penguins' caretaker, noticed the incubatory practice and gave Roy and Silo a fertile egg to tend. After thirty-four days, the couple successfully hatched "Tango" and cared for the chick until it could survive on her/his/its own. According to Smith, the caretaker was "full of praise for them" (B7).

Roy, Silo and Tango achieved celebrity status for some people, primarily those belonging to same-sex advocacy groups. The birds served as evidence, for these groups, of the naturalness of homosexuality (e.g., "If it is normal for animals to have same-sex attractions then it must be normal for humans"). Roy and Silo also served as evidence that same-sex human couples can effectively rear children (e.g., "If gay birds can successfully rear offspring then gay humans can too"). Roy, Silo and Tango started to function as media, as vehicles for promoting human social change. But a year after the penguin family made headlines a film was released that implicitly questioned the legitimacy of same-sex attracted penguins.

June 24, 2005. *March of the Penguins* begins airing in the United States. Narrated by Morgan Freeman, the film features a slice of penguin life as it occurs in Antarctica. Viewers catch a glimpse of a penguin reproductive cycle, the community ethos of bird hatching and rearing and the harsh living conditions that, based on human standards, penguins weather. The film becomes the highest grossing documentary after Michael Moore's *Fahrenheit 9–11* and stirs up just as much controversy.

September 13, 2005. Jonathan Miller describes how socially conservative groups appropriate penguins from *March of the Penguins* to promote monogamy, advance antiabortion arguments, display tenets of "intelligent design," and vilify homosexuality (Zuk). Pro-life activist Stanek enjoys the film because nearly "every scene and narrative verified the beauty of life and the rightness of protecting it." Stanek relates the *birds'* activities, depicted by the film's (human) creators, to her thoughts about *human* activities:

> As I watched [the movie], I wondered how people could ooh and aww when baby penguins pecked out of their shells, or cover their eyes when a giant petrel attacked a baby penguin, yet not give a thought to the dismemberment and killing of human babies.

For Stanek, the filmed penguins work as media, as vehicles for making arguments against human proabortion groups. And while the birds serve as conditions for Stanek's discourse—that is, while they make Stanek's claims possible—her arguments are not aimed at the birds.

September 20, 2005. In an editorial in the *Times* (London), Caitlin Moran responds to uses of penguins in *March of the Penguins*:

> Still in the US, a Christian audience is making a documentary about penguins the biggest factual cinema release since *Fahrenheit 9/11*. Churches are block booking seats for *March of the Penguins*. . . . To be honest, this is good news. If American Christians want to go public on the fact that they're now morally guided by penguins, at least we know where we all stand. (2.7)

Kluger also critiques the "religious right" for turning the "family film" into a "family feud," especially since Laura Kim, vice president of Warner Independent Pictures (the film's distributor), says, "'They're just birds'" (13A).

September 27, 2005. During the hype surrounding *March of the Penguins*, Roy and Silo emerge again, only this time in another human-created story: They separate. "A famous gay penguin at Central Park Zoo in New York has reignited the 'culture wars' over homosexuality by going straight," Bone writes in the *Times*:

> The widely publicized story of the two males bringing up a baby, named Tango, made them gay icons. Their same-sex household was cited by liberals as a corrective to the traditional "family values" displayed by Emperor penguins in the hugely popular new documentary, *The*[sic] *March of the Penguins*, which has been hailed by Christian Conservatives. But Silo walked out on Roy for a girlfriend from California called Scrappy, who moved to Central Park from SeaWorld in San Diego. The new heterosexual couple built a nest and hang out by the pool, while Roy broods alone.

Earlier the same week, FOXNews.com released an article subtitled "New York City's Most Famous Gay Penguin Couple Has Split Up." The spin FOX added to the story, however, was conservative when juxtaposed against the (liberal) *Times* article:

> Even worse, one of [the penguins] has taken up with a female penguin new to the Central Park Zoo . . . Silo and Roy, two male chinstrap penguins native to the South Atlantic, made local headlines six years ago when they came out with their same-sex relationship. Since then, the pair have successfully hatched and raised an adopted chick—after trying to incubate a rock—and become [sic] role models for six other same-sex couples among penguins at the zoo. . . . Roy, all alone, sat disconsolately at the edge of the penguin area, staring at the wall.

Throckmorton refers to Silo as the "world's first documented ex-gay penguin," and describes Scrappy as a "hot little bird" from San Diego. "I guess [Silo] was wishing for a California girl," Throckmorton writes.

> With Silo and Scrappy picking out curtains together, will gay rights groups now acknowledge that sexual orientation changes? The concept of gay penguin permanence painted by the [*New York*] *Times* and *And Tango Makes Three* [a children's book about Roy, Silo and Tango] now seems more like fiction than [a] public policy sign post.

While the *Times*, FOXNEWS.com and Throckmorton use penguins as media, they also "spin" Roy and Silo in particular ways. Roy and Silo become human-sexualized birds, and interpretations of their (lack of a) relationship differ with each source, that is, with the liberal *Times* or the conservative FOX and Throckmorton. Each promotes a liberal or conservative agenda directed at humans, but the birds themselves do not create this spin. Roy and Silo are also not the audience of these messages.

While debates about human same-sex marriage fueled debates about penguin sexuality, the straight penguins in *March of the Penguins* served as political fodder because they surfaced around the time that gay penguins surfaced in mainstream media outlets. The film aired without a "single queer bird to be found" (Uddin and Hobbes 1), and, combined with the credibility often attributed to documentaries, allowed the very presence of straight birds to function as antigay advocates. However, introducing gay animals into mainstream discourse assists progay groups that argue for the naturalness of human homosexuality and thus the need to protect and respect gay humans. But introducing gay animals into discourse can also perpetuate antigay rhetoric that deems gay beings "animalistic" and "uncivilized." With the penguins, homosexuality literally becomes "for the birds."

One event further dramatized the use of penguins in mass-mediated discourse: the separation of Roy and Silo. For antigay groups the separated birds served as evidence that human sexuality can "change" or "correct" itself. Roy and Silo's separation also perpetuated the unfaithful, promiscuous discourse commonly attributed to nonheterosexual humans, and Silo's "conversion" reinforced the heterosexual and thus factual story of *March of the Penguins* (that is, it could now be argued that gay penguins do not exist and therefore need not be in the film.) While penguins made the conditions for such discourse possible, the discourse was directed at humans, not the birds. The animals functioned as mediators of human communication.

Furthermore, in the penguin debates humans anthropomorphically sexualized the birds. This sexualizing, however, moved beyond anthropomorphism when the birds served as evidence, for humans, that "homosexuality is natural" or that "gay people can go straight" (i.e., since Silo went straight). Anthropomorphized penguins functioned as the necessary conditions for communication used by humans to target human populations with human-relevant messages. It is this idea that makes the penguins media.

Thus far, I have illustrated ways in which humans can use dogs, cats and penguins as media, as agencies to accomplish particular tasks as well as to advance particular goals, pleasures and ideologies. In the next section, I move from showing *how* humans have used animals as media to highlighting one area of concern about human animal-use. In particular, I discern possible consequences of human animal-use by criticizing the rhetoric of invasive species exhibits, rhetoric that calls for the eradication of harmful "illegal," "alien," "foreign" and "nonnative" animals and rhetoric that, when conceiving of animals as media, might promote hatred toward human populations labeled in such ways.

INVASIVE SPECIES AS MEDIA

Nonindigenous animals are "alien or exotic" beings that did not originate in a particular area (Lodge and Shrader-Frechette 32). These animals become "invasive" when they pose "significant threats to the environment, the economy and human health" (Olson 34), alter how ecosystems (should) function, work against biodiversity, contaminate the value of "regional biota" (Donlan and Martin 267) and affect the aesthetic desires of humans. Over seven thousand invasive animals live in the United States, and invasive species exhibits found at many zoos and aquaria provide strategies to stop "the introduction of new species" and to eradicate "existing [invasive] populations" (Swift 5). In 1999, Bill Clinton even developed the National Invasive Species Council, an institutionalized authority designed to protect native species.

Authorities categorize species as invasive depending on (human) aesthetic, temporal, environmental and economic goals. "What is harm for one person may be good for another," Lodge and Shrader-Frachette (32) write, further emphasizing the contingency of invasive categorization: "Any characterization that any or all nonindigenous species are good or bad is a value judgment, not science. In deciding which nonindigenous species are invasive, value judgments from many societal constituencies are required" (34–35). Lodge and Shrader-Frachette conceive of invasive species as natural and believe definitions of "nonindigenous" and "invasive" hinge upon the "time frame" being considered (33); human temporal "benchmarks dictate who is native and who is not" (Donlan and Martin). Thus, animals that arrived in a region centuries ago may easily acquire a "native" status whereas those that arrived within the last few decades may not get such a privilege.

In order to demonstrate how the rhetoric of invasive species can be harmful when viewing animals as media, I must acknowledge two assumptions. First, I assume that museum, zoo and aquarium exhibits are "partial" and "partisan" (Goodall 55). As Bernard Armada suggests, "By privileging certain narratives and artifacts over others, museums implicitly communicate who/what is central and who/what is peripheral; who/what we must remember and who/what it is okay to forget" (236). Karp argues that museum exhibits are "bound up with assertions about what is central or peripheral, valued or useless, known or to be discovered, essential to identity or marginal" (7; see also Dickinson, Ott, and Aoki; Haraway, "Teddy Bear"). Invasive exhibits categorize invasive animals as peripheral, useless, marginal and okay-to-forget beings.

Second, I assume that language is both representational *and* constitutive—the words we use to describe a phenomenon influence our understanding and evaluation of this phenomenon *and* call this phenomenon into existence (Althusser). "The seeing created by language," Ricoeur writes, is "not a seeing of this or that; it is a 'seeing-as'" (133). And language use—naming—is a seeing-as "permeated with attitudes, opinions, and social evaluations" (Ellis 215; Lakoff and Johnson), a seeing-as that constitutes identities, "social relations" and "systems of knowledge and belief" (Butler; Fairclough 309; Foster; Milstein), and a seeing-as that matters.

These two assumptions coupled with the animals-as-media theory make the use of animals in invasive exhibits especially salient, prompting such questions as "What kinds of human sensemaking do invasive animals make possible?" and, more specifically, "What ideologies do invasive animals harbor, ideologies that implicate not only the animals themselves but also human populations?"

Furthermore, I previously noted that when humans use animals to speak to other humans, the discourse is not often directed at the animals. With invasive species discourse, however, this is inaccurate. Invasive discourse implicates the animals discoursed about—the goal of invasive discourse is to eradicate the harmful, nonnative, invasive beings.

When invasive animals function as media, they enable harmful kinds of human sensemaking. For instance, the New York State Zoo (Thompson Park) calls invasive animals "alien invaders." The Point Defiance Zoo and Aquarium (Tacoma, WA) refers to invasives as "bad neighbors" (Olson 35). And Shedd Aquarium (Chicago, IL) defines invasives as "animals that arrive in a place where they didn't originate, then multiply, spread and do harm in their new environment" (Holland and Mason 1). In a study of the Florida Aquarium in Tampa, Serrell & Associates find that visitors often refer to invasive animals as nonnative beings who "take over" natives, beings who invade "the area of another" and beings who "harm" and "destroy" and do not "belong here" (1; see Dale; Reed; Zayas). Serrell & Associates argue, however, that invasive animals should not be described as "'foreign invaders'" because these words may suggest a "double meaning post 9/11" (4). Moreover, as a former volunteer at the Florida Aquarium, I was asked by authorities to refrain from calling invasive animals "aliens" and "nonnatives" because these terms resemble the language of human citizenship.

These examples illustrate that the ways invasive animals are taken up in discourse—that is, the ways they are languaged—resemble the ways that human "immigrants" are taken up in discourse. In other words, animals in invasive exhibits—exhibits designed to eradicate harmful "nonnative," "alien" and "foreign" beings—are described in terms similar to those used to describe human immigrants. And this matters because a dislike of nonnatives, aliens and foreigners—or of phenomena labeled as such—becomes institutionally justified by animals in invasive exhibits. By advocating for linguistic caution, Serrell & Associates and the Florida Aquarium authorities realized that language matters; that animals and the terms used to describe animals enable particular kinds of human sensemaking. They realized that animals could function as media.

I also call attention to the ethnic identifiers of invasive animals, such as "Asian" carp, "Norway" rats, "Japanese" Shore Crabs. When animals are labeled using ethnicized or nationalized terms and then placed in exhibits that promote the eradication of ethnicized animals, the conditions develop for a devaluing of similarly labeled humans. In other words, if language matters, then a desire to eradicate Asian animals has relevance for Asian-labeled humans, especially when the language used to name undesired animals is the same language used to name humans. I do not suggest that when people see a headline such as "Chicago Canal Flooded with Toxin to Kill Asian Carp" (Schaper) they unquestioningly want to kill anything Asian. However, I do suggest that negative attitudes toward Asians are enabled and perpetuated by the inclusion of animals in anti-Asian discourse. Reconsider Serrell & Associates' concern about calling invasives "illegal aliens." Why? Because alien animals can mediate negative sentiments between these animals and alien humans; as Gergen observes, "patterns of action are typically intertwined with modes of discourse" (115).

If language is both representative and constitutive and if animals can function as media, then invasive animals enable an insidious form of anti-immigration discourse found in prominent cultural institutions. Again, I do not think that animals in invasive exhibits "inject" people with nativist beliefs or that people will unwittingly embrace xenophobic attitudes. I am also not questioning the legitimacy of invasive exhibits. I am only calling attention to the kinds of sensemaking that invasive animal discourse—discourse created by humans for other humans—makes possible. Conceiving of animals as media makes for a new way of understanding such taken-for-granted, institutional discourse.

ANIMALS AS MEDIA: CONSEQUENCES AND POSSIBILITIES

Throughout this chapter, I have demonstrated ways humans can use animals for human purposes. I have also discerned reasons why an institution should refrain from using particular animals (e.g., invasive species) in particular ways (e.g., to promote anti-immigration discourse). In this last section, I use tenets of symbolic interaction to further theorize human animal-use. I offer a general discussion of how animals can function in (human) discourse and note why we cannot take animals—and human uses of animals—for granted.

Symbolic interactionists highlight how meanings emerge in interaction and illustrate how humans use these meanings to make sense of themselves, others and society. Thus, in this section, I describe how animals can influence meaning-making processes. I specifically illustrate how animals function as symbols of "flexible innocence," symbols that can come to mean a variety of things and symbols that can seem free of intention and bias. Using examples from my three cases, I describe how the concept of flexible innocence develops and conclude by demonstrating what the concept suggests about personal accountability in human interaction.

Animals can function as flexible symbols in human interaction. They can appear happy or sad, cold or warm, gay or straight, as having acted with malicious intent or having not a care in the world. Such ascriptions, however, can only happen if *a human puts forth appropriate evidence.*

Since animals do not speak the same languages as humans, they cannot dispute what humans say about them, contest how they are represented in language or challenge the human use of them as evidence. As ambiguous entities, animals can thus be molded to fit human discourse in a plethora of ways, all of which remain difficult to discern, disprove and critique.

For instance, consider the statements "'Stop fighting—you're upsetting the dog'" and "'Get quiet—the dog is upset'" (Cain 76). While many scholars acknowledge the ability of humans and animals to communicate (Arluke and Sanders; Irvine; Haraway), I argue that a human can never know, with any certainty, what a dog thinks or feels; she or he can only infer *in human terms*. An animal is "strategically ambiguous" (Eisenberg) in that it can be made to mean a variety of things; a look of disgust could resemble a look of fear, sickness or apathy depending on the context, interpretations and arguments offered by humans.

The flexibility of animals in interaction can also be observed with the gay and straight penguins. No person or group could definitively know what the penguins were doing or what their actions meant; they came to mean a variety of things for a variety of constituencies. For instance, the story of Roy and Silo emphasized the naturalness of human homosexuality; however, it also framed homosexuality as animalistic, uncivilized and subject to change. The penguins in *March of the Penguins* came to suggest that homosexuality did not exist in the wild—that is, outside of zoos and aquaria; however, the existence of gay penguins in zoos and aquaria disqualified the accuracy of the film.

Animals can function as "innocently involved" symbols in human interaction as well (Goffman, "Strategic"). This innocence can develop in two ways. First, when animals are conceived of as politically neutral beings, as creatures believed to be without intention and bias, then human uses of animals can—because of the animal "content"—make such use seem innocent, depoliticized and indisputable. For instance, some reviewers of earlier versions of this essay have criticized me for criticizing the language of invasive species. They responded by saying that invasive animals—and the language affiliated with the animals—were innocent, scientific and thus bias free; it did not matter how they were labeled.

The ways in which media are understood can also add a layer of disinterest to ways meaning happens in human interaction. Media are often assumed to have little influence on these meanings (McLuhan; Ong; Schivelbush). Thus, when animals work as media, it can be difficult to acknowledge their influence on the interaction. Laura Kim's reference to the penguins in *March of the Penguins* as "'just birds'" is indicative of this disinterested logic. The penguins were not relevant to and are disinterested in human affairs. However, such an ascription of disinterest serves a purpose: By perceiving an animal as disinterested, humans can easily disregard the human ideologies the animal harbors. This disregard can allow the animal to come across as an innocent and thus meaningless variable in (human) interaction.

When an animal comes to possess an aura of flexible innocence, the human origins of animal discourse become distanced from the human originator(s). As such, the human use of the animal can function, for the human, as an "avoidance ritual" (Goffman, "Interaction"), an act that can make the human less "accountable" (Goffman, "Strategic"; Scott and Lyman) for what is said or done with the animal. The human origins of animal discourse become "hinted" origins, ones that can be disregarded and denied (Goffman, "Interaction" 30), and any ideas of human strategy or use of the animals become futile; because of the aura of innocence, the human use of animals, consequently, can be seen as innocent, too.

For instance, it would be politically incorrect for someone, especially a public figure, to say, explicitly, that a foreigner, an illegal alien, a nonnative human should be eradicated. However, by using a flexible and innocent animal as a medium, it is easy for a human to say that a foreign, alien, nonnative animal should be eradicated. The human messenger can deny responsibility for her or his role in the (offensive) claim by saying that science supports it or that the message originated with the invasive animal or that such a message has no relevance for human populations labeled "illegal," "alien," "foreign" or "nonnative." By using an animal to avoid personal accountability, a human can engage in "tactful blindness," a face-saving technique (Goffman, "Interaction" 18), especially since the animal cannot talk back.

An area of this research that needs to be further addressed is how animals are implicated by the messages they mediate. For example, the discourse made possible by and taken up in television does not implicate television itself; for example, a television news reporter is able to be a television news reporter partially because of the television, but what the reporter says does not implicate television itself. However, the discourse made possible and taken up by an animal often does implicate the animal: Humans use dogs, penguins and invasive species to speak to other humans, but *what* humans say often implicates the animals used to say this *what*; saying a "foreign" animal should be killed will have consequences for this animal *as well as* suggest to other humans ways in which foreigners should be treated.

Conceiving of animals as media means that we must attend to the ways in which animals function in human relations. This means attending to the ways humans talk to and about animals, the ways humans use animals in television programs, films and advertisements and the ways aquariums and zoos represent particular species. Animals *are* important interactional phenomena, especially when they produce and perpetuate harmful human agendas; we must not take them, or their affiliated humans, for granted.

REFERENCES

Adam, Barry D. "Love and Sex in Constructing Identity among Men Who Have Sex with Men." *International Journal of Sexuality and Gender Studies* 5.4 (2000): 325–339. Print.

Althusser, Louis. "Ideology and Ideological State Apparatuses (Notes Towards an Investigation)." *Lenin and Philosophy and Other Essays.* Trans. Ben Brewster. London: NLB, 1971. 121–173. Print.

Anderson, David C. "The Human-Companion Animal Bond." *The Reference Librarian* 86 (2004): 7–23. Print.

Arluke, Arnold, and Clinton R. Sanders. *Regarding Animals.* Philadelphia: Temple UP, 1996. Print.

Armada, Bernard J. "Memorial Agon: An Interpretive Tour of the National Civil Rights Museum." *Southern Communication Journal* 63.3 (1998): 235–43. Print.

Bagemihl, Bruce. *Biological Exuberance: Animal Homosexuality and Natural Diversity.* New York: St. Martin's P, 1999. Print.

Berger, John. *About Looking.* New York: Pantheon, 1992. Print.

Bone, James. "Gay Icon Causes a Flap by Picking Up a Female." *Times* (London) 27 Sept. 2005. Web. 10 Dec. 2005.

Butler, Judith. *Excitable Speech: A Politics of the Performative.* New York: Routledge, 1997. Print.

Cain, Ann Ottney. "A Study of Pets in the Family System." *New Perspectives on Our Lives with Companion Animals.* Ed. Aaron Honori Katcher and Alan M. Beck. Philadelphia: U of Pennsylvania P, 1983. 72–81. Print.

Cardoze, Cristina. "They're in Love. They're Gay. They're Penguins . . . and They're Not Alone." *Columbia News Service* 10 Jun. 2002. Web. 17 Oct. 2005.

Challinor, David. "Introduction: Contrasting Viewpoints." *Perceptions of Animals in American Culture.* Ed. R. J. Hoage. Washington, D.C.: Smithsonian Institution P, 1989. 1–24. Print.

Charmaz, Kathy. "The Grounded Theory Method: An Explication and Interpretation." *Contemporary Field Research: A Collection of Readings.* Ed. Robert M. Emerson. Prospect Heights, IL: Waveland, 1983. 109–125. Print.

Chris, Cynthia. *Watching Wildlife.* Minneapolis: U of Minnesota P, 2006. Print.

Creswell, John W. *Research Design: Qualitative, Quantitative, and Mixed Method Approaches.* 3rd ed. Thousand Oaks, CA: Sage, 2008. Print.

Dale, Kevin. "Community Considers How to Get Rid of Unwanted Iguanas." *Tampa Tribune* 8 Feb. 2006: 4. Print.

Denzin, Norman K. *Interpretive Interactionism.* Newbury Park, CA: Sage, 1989. Print.

Desmond, Jane C. *Staging Tourism: Bodies on Display from Waikiki to Sea World.* Chicago: U of Chicago P, 1999. Print.

Dickinson, Greg, Brian L. Ott, and Eric Aoki. "Memory and Myth at the Buffalo Bill Museum." *Western Journal of Communication* 69.2 (2005): 85–108. Print.

Donlan, C. Josh, and Paul S. Martin. "Role of Ecological History in Invasive Species Management and Conservation." *Conservation Biology* 18.1 (2004): 267–269. Print.

Eisenberg, Eric. *Strategic Ambiguities: Essays on Communication, Organization, and Identity.* Thousand Oaks, CA: Sage, 2007. Print.

Ellis, Donald G. "Post-Structuralism and Language: Non-Sense." *Communication Monographs* 58.2 (1991): 213–224. Print.

Evans, Andrew. "World's First Exgay Penguin?" *Exgaywatch.com* 21 Sept. 2005. Web. 20 Nov. 2006.

Fairclough, Norman. "Critical Analysis of Media Discourse." *Media Studies: A Reader.* Ed. Paul Marris and Sue Thornham. New York: New York UP, 2000. 308–328. Print.

Fiske, John. "Medium/Media." *Key Concepts in Communication and Cultural Studies.* Ed. Tim O'Sullivan et al. New York: Routledge, 1994. 176–177. Print.

Foster, Elissa. "Commitment, Communication, and Contending with Heteronormativity: An Invitation to Greater Reflexivity in Interpersonal Research." *Southern Communication Journal* 73.1 (2008): 84–101. Print.

FOXNews.com. "Gay Penguins Break Up." 16 Sept. 2005. Web. 22 Apr. 2010.

Foucault, Michel. *The History of Sexuality.* Trans. Robert Hurley. Vol. 1. New York: Vintage, 1978. Print.

Francione, Gary L. "Animals—Property or Persons?" *Animal Rights: Current Debates and New Directions.* Ed. Cass R. Sunstein and Martha C. Nussbaum. New York: Oxford UP, 2004. 108–142. Print.

Gergen, Kenneth J. *Realities and Relationships: Soundings in Social Construction.* Cambridge, MA: Harvard UP, 1997. Print.

Giffney, Noreen, and Myra J. Hird. *Queering the Non/Human.* Burlington, VT: Ashgate, 2008. Print.

Goffman, Erving. *Interaction Ritual: Essays on Face-to-Face Behavior.* Garden City, NY: Anchor, 1967. Print.

———. "Strategic Interaction." *Strategic Interaction.* Philadelphia: U of Pennsylvania P, 1969. 85–145. Print.

Goodall, H. L. *Writing the New Ethnography.* Walnut Creek, CA: AltaMira P, 2001. Print.

Hallett, Vicki. "Make Way for Gay Penguins." *U.S. News & World Report* 28 May 2005. Web. 16 Dec. 2006.

Haraway, Donna. "Teddy Bear Patriarchy: Taxidermy in the Garden of Eden, New York City, 1908–1936." *Primate Visions: Gender, Race and Nature in the World of Modern Science.* New York: Routledge, 1989. 26–58. Print.

———. *When Species Meet.* Minneapolis: U of Minnesota P, 2008. Print.

Holland, M., and E. Mason. "Asian Carp, Snakehead and Gobies Top List at First-Ever Invasive Species Exhibit at Shedd Aquarium." *Shedd Aquarium.* 5 Jan. 2006. Web. 14 Apr. 2006.

Irvine, Leslie. "A Model of Animal Selfhood: Expanding Interactionist Possibilities." *Symbolic Interaction* 27.1 (2004): 3–21. Print.

March of the Penguins. Dir. Luc Jacquet. Perf. Morgan Freeman. Warner Independent Pictures, 2005. Film.

Jerolmack, Colin. "Our Animals, Our Selves? Chipping Away the Human-Animal Divide." *Sociological Forum* 20.4 (2005): 651–660. Print.

Karp, Ivan. "Introduction: Museums and Communities: The Politics of Public Culture." *Museums and Communities: The Politics of Public Culture.* Ed. Ivan Karp, Christine Mullen Kreamer, and Steven D. Lavine. Washington: Smithsonian Institution P, 1992. 1–17. Print.

King, Margaret J. "The Audience in the Wilderness: The Disney Nature Films." *Journal of Popular Film and Television* 24.2 (1996): 60–68. Print.

Kluger, Bruce. "Latching On to Us vs. Them." *USA Today* 8 Nov. 2005: 13A. Print.

Knight, Robert. "Wedding March of the Penguins." *Worldnetdaily.com* 22 Sept. 2005. Web. 13 Nov. 2006.

Lakoff, George, and Mark Johnson. *Metaphors We Live By.* Chicago: U of Chicago P, 1980. Print.

Levine, K. "Santa, Help Us Pets Train Our Owners to Stay Trendy." *The Tampa Tribune* 25 Nov. 2006: 12. Print.

Liddiard, Mark. "Changing Histories: Museums, Sexuality and the Future of the Past." *Museum and Society* 2.1 (2004): 15–29. Print.

Lodge, David M., and Kristin Shrader-Frechette. "Nonindigenous Species: Ecological Explanation, Environmental Ethics, and Public Policy." *Conservation Biology* 17.1 (2003): 31–37. Print.

Lorde, Audre. "Age, Race, Class, and Sex: Women Redefining Difference." *Sister Outsider.* Berkeley, CA: The Crossing Press, 1984. 114–123. Print.

McLuhan, Marshall. *Understanding Media: The Extensions of Man.* New York: McGraw-Hill, 1964. Print.

Mead, George Herbert. *Mind, Self, and Society from the Standpoint of a Social Behaviorist.* Chicago: U of Chicago P, 1962. Print.

Messent, Peter R. "Social Facilitation of Contact with Other People by Pet Dogs." *New Perspectives on Our Lives with Companion Animals.* Ed. Aaron Honori Katcher and Alan M. Beck. Philadelphia: U of Pennsylvania P, 1983. 37–46. Print.

Meyrowitz, Joshua. *No Sense of Place: The Impact of Electronic Media on Social Behavior.* New York: Oxford UP, 1985. Print.

Middlesexes: Redefining He and She. Dir. Anthony Thomas. Perf. Gore Vidal. HBO, 2005. Film.

Miller, Jonathan. "March of the Conservatives: Penguin Film as Political Fodder." *New York Times* 13 Sept. 2005: F2. Print.

Miller, Kenneth. "Meet the Doodles." *Life* 8 Oct. 2004: 15–17. Print.

Milstein, Tema. "Nature Identification: The Power of Pointing and Naming." *Environmental Communication* 5.1 (2011): 3–24. Print.

Mooallem, Jon. "Can Animals Be Gay?" *The New York Times Magazine* 29 Mar. 2010. Web. 22 Apr. 2010.

Moran, Caitlin. "Penguins Lead Way." *Times* (London) 20 Sept. 2005: 2.7. Print.

Olson, Steve. "Aquatic Invasive Species." *Communiqué* Nov. 2004: 34–35. Print.

Ong, Walter J. *Orality and Literacy: The Technologizing of the Word.* New York: Routledge, 1982. Print.

Out in Nature. Dir. Stéphanie Alexandresco, Bertrand Loyer, and Jessica Menendez. New York: Logo, 2001. Television.

Pauly, John J. "Media Structures and the Dialogue of Democracy." *Dialogue: Theorizing Difference in Communication Studies.* Ed. Rob Anderson, Leslie A. Baxter, and Kenneth N. Cissna. Thousand Oaks, CA: Sage, 2004. 243–58. Print.

Prager, Emily. "Fashion Unleashed." *New York Times* 20 Feb. 1994, sec. 9: V1–V8. Print.

Ramirez, Michael. "'My Dog's Just like Me': Dog Ownership as a Gender Display." *Symbolic Interaction* 29.3 (2006): 373–391. Print.

Reed, Matt. "Non-Native Iguanas Sink Their Claws into Fla. Island." *USA Today* 27 Jun. 2006: 3A. Print.

Reiss, Michael J. "Imagining the World: The Significance of Religious Worldviews for Science Education." *Science & Education* 18 (2009): 783–796. Print.

Richardson, Justin, and Peter Parnell. *And Tango Makes Three.* New York: Simon and Schuster, 2005. Print.

Ricoeur, Paul. "The Function of Fiction in Shaping Reality." *Man and World* 12.2 (1979): 123–141. Print.

Robins, Douglas M., Clinton R. Sanders, and Spencer E. Cahill. "Dogs and Their People: Pet-Facilitated Interaction in a Public Setting." *Journal of Contemporary Ethnography* 20.1 (1991): 3–25. Print.

Roughgarden, Joan. *Evolution's Rainbow: Diversity, Gender, and Sexuality in Nature and People.* Berkeley: U of California P, 2004. Print.

Sanders, Clinton R. "'The Dog You Deserve': Ambivalence in the K-9 Officer/Patrol Dog Relationship." *Journal of Contemporary Ethnography* 35.2 (2006): 148–172. Print.

Schaper, David. "Chicago Canal Flooded with Toxin to Kill Asian Carp." *NPR.org.* 4 Dec. 2009. Web. 6 Mar. 2012.

Schivelbusch, Wolfgang. *The Railway Journey: The Industrialization of Time and Space in the 19th Century.* Berkeley: U of California P, 1986. Print.

Scott, Marvin B., and Stanford M. Lyman. "Accounts." *American Sociological Review* 33.1 (1968): 46–62. Print.

Serrell & Associates. "A Front-End Evaluation on Invasive Species for the Florida Aquarium." *Informalscience.org.* 28 Nov. 2001. Web. 10 Apr. 2006.

Smith, Dinitia. "Love That Dare Not Squeak Its Name." *New York Times* 7 Feb. 2004: B7. Print.

Spigel, Lynn. "Communicating with the Dead: Elvis as Medium." *camera obscura* 23 (1991): 177–205. Print.

Stake, Robert E. "Qualitative Case Studies." *Handbook of Qualitative Research.* Ed. Norman K. Denzin and Yvonna L. Lincoln. 3rd ed. Thousand Oaks, CA: Sage P, 2005. 443–466. Print.

Stanek, Jill. "'March of the Penguins' vs. March for Women's Lives." *Worldnetdaily.com* 10 Aug. 2005. Web. 11 Dec. 2005.

Swift, Earl. "Can They Be Stopped?" *Parade* 22 May 2005: 4–5. Print.

Tapper, Jake. "Can We Judge a Man by His Dog?" *Tampa Bay Times* 13 Jul. 2007: 12. Print.

Tapper, Jake, and Sunlen Miller. "Did Obama Break a Pledge to Adopt a Rescue Dog?" *ABC News.* 13 Apr. 2009. Web. 22 Apr. 2010.

Terry, Jennifer. "'Unnatural Acts' in Nature: The Scientific Fascination with Queer Animals." *GLQ: A Journal of Lesbian and Gay Studies* 6.2 (2000): 151–193. Print.

Throckmorton, Warren. "Silo Rains on the Penguin Pride Parade." *Catholiceducation.org.* 19 Sept. 2005. Web. 20 Nov. 2006.

The Truth about Gay Animals. Dir. H. Littleboy. Exec. Prod. N. Minhas. Los Angeles: Trio, 2002. Television.

Uddin, Lisa, and Peter Hobbs. "Introduction: Nature Loving." *Invisible Culture.* 2005. Web. 15 Sept. 2005. 9.

Wagenseil, Paul. "Gay Penguins Break Up." *FOXNews.com* 16 Sept. 2005. Web. 17 Oct. 2005.

Williams, David. "Inappropriate(d) Others or, the Difficulty of Being a Dog." *TDR: The Drama Review* 51.1 (2007): 92–118. Print.

Yin, Robert K. *Case Study Research: Design and Methods.* Thousand Oaks, CA: Sage, 1994. Print.

Zayas, Alexandra. "'Evil Plant' Invades, Takes Over." *Tampa Bay Times* 23 May 2008: 13. Print.

Zuk, Marlene. "Family Values in Black and White." *Nature* 23 Feb. 2006: 917. Print.

3 Beached Whales
Tracing the Rhetorical Force of Extraordinary Material Articulations

Deborah Cox Callister

INTRODUCTION

Rhetorical theory has long been plagued by a material/discursive binary. This chapter explores linkages between these two concepts and ways in which they are mutually constitutive. Marxist feminist theorist Dana Cloud treats them as a dyad, arguing that governmental apparati, institutional structures and taken-for-granted daily practices must be accounted for, because they present material barriers for people who experience social and economic marginalization. This critique charges that rhetorical theory erases the material by apprehending rhetoric as text, thereby privileging discursive constructions of realities. Environmental communication scholar Richard Rogers moves the discussion beyond humans, explaining that the marginalization and objectification of nature is a result of the anthropocentric ontological and epistemological assumptions that constitute communication theory. This critique presents an even larger set of tensions.

For example, how can a discipline that apprehends everything as text account for communication between humans and the natural world, except through the discursive meanings ascribed to it by human rhetors, critics and audiences? Although I agree with environmental ethicist Bryan Norton that perhaps the most humans can achieve is some form of weak anthropocentrism (because we are constrained by our own corporeality), I suggest that focusing on extraordinary material events might be one way for rhetorical theory to begin to account for and deal with animals and nature. For example, whale beachings, as extrahuman material ruptures, are a window into understanding two different ideologies about animals and nature in hegemonic and counterhegemonic discourses. Moreover, we can draw from Ronald Greene's materialist rhetorical theory and Kevin DeLuca's definition of articulation in order to respond to environmental communication scholar Richard Rogers's call for decentering humans in constitutive theories toward a transhuman dialogue with the natural world.

A dispute over the Navy's use of active sonar in antisubmarine training provides the context for this analysis. In 2007, campaigns to save the whales from deleterious harm associated with low frequency sonar testing off the coast of Southern California resulted in a court injunction that temporarily halted the sonar testing. The Navy appealed the decision, and on November 13, 2008, Bob Egelko, a *San Francisco Chronicle* staff writer, framed the U.S. Supreme Court's decision as follows: "Threats to national security are more important than possible harm to whales and dolphins" (n. pag.).

It is within this context that I explore how whales' bodies on beaches can act as a flashpoint whereby humans with disparate ideologies about animals and nature deploy divergent rhetorics. I draw on environmental communication scholar Kevin DeLuca's definition of articulation as the "act of linking" elements or floating signifiers, in such a manner "that they can be understood as being spoken anew" (*Image Politics* 38). This conception of articulation carries rhetorical force. Drawing from rhetorical materialist Ronald Greene's theory of rhetoric as articulation and deliberative practice, I then identify and trace the deployment of whale rhetorics that emanate from this material articulation or the linkage of whale bodies with beaches onto a field of hegemonic and counterhegemonic discourses that produce disparate narratives. Critical cultural theorist Stuart Hall defines hegemony as "a form of power based on leadership by a group in many fields of activity at once, so that its ascendancy commands widespread consent and appears natural and inevitable" (259). Thus, hegemonic discourses prop up the status quo and counterhegemonic discourses are those that oppose these dominant forms of power. Finally, I analyze counterhegemonic and hegemonic narratives that are deliberated by publics and explore their consequences for acoustically sensitive marine mammals. This mapping of connections to antithetical discourses, each with unique material consequences, builds on Greene's materialist theory of rhetoric by using the logics of both representation (e.g., media) and articulation to gain an understanding of ways in which a government can program reality. It also responds to Rogers's call for rhetoric to deal more extensively with human/nature interactions by demonstrating one way to accomplish this work.

ARTICULATING CONSTITUTIVE, CRITICAL AND MATERIAL RHETORIC

Traditionally, rhetorical criticism analyzes persuasion in texts such as public speeches (e.g., Simons). In the last few decades of the twentieth century, several turns took place in the academy that altered the landscape of inquiry. For example, in the ideological and critical turns, we can begin to see rhetorical theory and criticism account for material structures, corporeal experiences and power through social and symbolic practices within a human domain. These developments have helped us to understand the ways in which realities are socially constructed, but our ability to account for human relationships with the natural world and contemporary environmental exigencies remains wanting. Let me explain.

In the ideological turn, rhetorical theorists Michael Calvin McGee and Philip Wander open rhetorical theory to the analysis of the ideological meanings with which texts are laden (McGee, "Ideological Turn"; Wander, "Third Persona"). These developments are followed by closer examinations of the role of power and a subsequent critical turn in the academy. For instance, critical theorist Raymie McKerrow invoked Michel Foucault's notion of power to argue that oppressive power is ubiquitous, requiring a critical rhetoric or an ongoing critique of truth claims produced through dominant discourses along with a simultaneous critique of freedom. Rhetorical and feminist theorist Barbara Biesecker contributes to this development of *discursive* conceptualizations of power and resistance. She argues that McKerrow's foregrounding of oppressive power leads to an oversight regarding Foucault's more developed theory of resistance. Biesecker asserts that "to understand power only as oppressive is reductive"; power relations, then, must be distinguished from "concrete discourse/practices of domination operative within specific regimes" (354). Thus, state apparati and other material institutions can be thought of as the terminal forms that power takes. In this way individuals, both within dominant and nondominant structures and among speech communities, mobilize power.

Dana Cloud challenges rhetoricians to look beyond discursive constructions of material structures toward their effects created through capitalist production and distribution of goods. Cloud argues for "a materialist theory of communication that recognizes the importance of material forces (economic and physical) in relation to rhetorical action" (144). This critique recognizes a materialist base and an idealist superstructure, and yet it remains bound within a human domain. Ronald Greene takes up the material/discursive binary and theorizes a different materialist rhetoric. He suggests that critics should focus on "how rhetorical practices create the conditions of possibility for a governing apparatus to judge and program reality" (22). In other words, Green's conception of articulation entails the nodes where "a set of human technologies come together to form a governing apparatus . . . that [theoretically] improves the welfare of a population" (32). Whereas Greene suggests that "a logic of representation" should be replaced by "a logic of articulation," I suggest that we need *both* in order to understand "how rhetoric makes possible the ability to judge and plan reality in order to police a population" (21). In fact, we can begin to identify and analyze how the logics of articulation and representation affect not just human but extrahuman populations.

This theoretical development is crucial to my interest in developing ways for rhetorical criticism to address the natural world. A logic of articulation enables us to move beyond the false binary of discourse versus the material, because it enables an understanding of rhetorical forces created by linkages that are simultaneously material and discursive. For example, as I will explain further below, rhetorical appeals deployed from the bodies of beached or stranded whales, and aimed at preventing or justifying more whale beachings, are loci where articulation is constituted by both material *and* discursive elements. This, however, does not fully account for the set of tensions presented to rhetoricians by the natural world.

Richard Rogers begins to address these tensions by criticizing communication theory in general, because it is defined not only through its ontological notions of discursivity, but through its resultant erasure and objectification of what he calls the "natural" material. Rogers asserts, "Constitutive theories follow idealism in treating the realm of the non-linguistic (matter, nature, 'reality') as inert and insignificant—in need of ordering via discourse" (246). In this manner, Rogers challenges communication scholars to develop theories that account for human relationships with the natural world. Kevin DeLuca advances the project by featuring materiality in his analysis of environmental activism and the radical juxtaposition of human bodies with nature as unruly arguments or antagonisms to hegemonic industrial discourse ("Unruly Arguments"). In fact, DeLuca defines articulation as having "two aspects: speaking forth elements and linking elements. Though elements preexist articulations as floating signifiers, the act of linking in a particular articulation modifies their character such that they can be understood as being spoken anew" (*Image Politics* 38). Certainly DeLuca's description of human beings buried up to their necks on logging roads to protest the passage of logging trucks in North America is a vivid example of articulation, because human bodies and the earth are linked or articulated in such a manner that they speak with tremendous rhetorical force (*Image Politics* 125).

This node of articulation serves as a flash point that deploys images of vulnerable bodies embedded within a vulnerable ecosystem, in North Kalmiopsis, Oregon. These rhetorical images intersect with hegemonic and counterhegemonic discourses that represent what Green (1998) calls a "cartography of deliberative rhetoric" that distributes "discourses, institutions and populations onto a field of action" (22). Moreover, what DeLuca calls "guerrilla imagefare in the woods" entails extraordinary material articulation between human bodies and soil (*Image Politics* 5). Thus, articulation as "speaking forth and linking elements" can be conceptualized along a continuum from solely discursive elements (e.g., connecting words to form a newly coined term such as " guerilla imagefare") to a blend of discursive and material elements (e.g., the symbolic meanings deployed from the site of vulnerable humans bodies physically buried in soil) to wholly material elements (e.g. the corporeal linkage of human bodies with soil). Perhaps proximity of discourse to the loci of wholly material articulations is a more fruitful way to think about this false discursive/material dyad. In other words, to a degree, discursive and material elements are mutually constitutive. Let me explain.

I have suggested that we can draw from Greene's materialist theory and DeLuca's definition of articulation as one way to respond to Rogers's call for decentering humans in constitutive theories toward a transhuman dialogue with the natural world. By focusing on extraordinary nonhuman articulations (in this case whale bodies articulating with beaches) we can explore human/nature relationships using Greene's dynamic interplay of deliberative rhetorics put in motion by these nonhuman whale subjects and their massive bodies that have evolved to float, communicate and navigate in seawater. We might think of discursive representation of human/beached whale encounters in terms of the degree to which they have traveled from the point of solely material articulation.

First, the physical encounter of whale/human bodies involves an extraordinary material articulation—the linking of two surprising material elements. Just as whales don't typically occupy beaches, humans don't typically encounter whales on beaches. Each of these distinct events involves unusual material linkages and, I posit, a momentary flash wherein such articulations remain wholly material and in the realm of corporeal experience (i.e., extradiscursive at least from a human perspective). But, the extraordinary material articulation of whales with beaches, and humans' encounters with the same, *speaks* with tremendous rhetorical force, and beached whale rhetorics get deployed from the bodies of these distressed or deceased whales. Disparate rhetorics connect to fields of discourses and discursive formations across institutions, media and daily human conversations. Through these discursive practices, rhetorical appeals are invented that undergo deliberation in the public sphere and in branches of government. In this way beached whale rhetorics, deployed from the site of material articulation, can be conceived as material and discursive imbrications (Cox 12). And yet, as these deployments of rhetorics reverberate through discursive formations in the form of narratives carrying symbolic codes as they travel, the material articulation or point of origin grows more distal, as do the corporeal experiences that they *originally* referenced.[1]

Disparate ideologies result in conflicting narratives that create controversies. As dominant and alternative ideologies collide in the public sphere, innovative rhetorical appeals create the conditions for governmental intervention that regulate practices. For example, in the case of whale beachings that took place in areas adjacent to naval antisubmarine sonar testing during the George W. Bush administration, whale activists clamored to hold the Navy accountable for specific whale beachings and eventually won a court injunction to stop the sonar testing off the coast of Southern California, but attorneys representing naval-training interests strove to elide any connection between sonar testing and whale beachings. Eventually discourses of homeland security trumped environmental discourses in a U.S. Supreme Court decision, but the deliberative rhetorics that ensued within the courts at least pushed the government toward recognition of connections between whale beachings and antisubmarine sonar testing and altered the naval sonar testing practices to mitigate their effects on whales (Swinford). However, the high court ruling, in the name of national security, creates an exception for the Navy under the administrative branch of government to literally and circuitously write dominant discourses back onto the bodies of acoustically sensitive cetaceans in the form of active sonar.

In sum, by explicating above and demonstrating below how this occurs, I am extending Greene's call to analyze "how rhetorical practices create the conditions of possibility for a governing apparatus to judge and program reality" beyond the human domain (22). Second, this essay deals with the natural world because it focuses on extrahuman material articulations (in this case beached whales) as a window into understanding two different ideologies about nature and animals in hegemonic and counterhegemonic discourses. Third, we can explore dominant and resistive forms of power through this model and begin to hold rhetorical practices (or habits) more accountable for their extratextual consequences, including their "natural" material effects. In short, analyzing extraordinary material articulations might be one way for rhetorical criticism to deal with the anthropogenic pressures that are pushing the carrying capacity of the earth toward the inability to sustain natural living systems.

DISPLACED WHALES AND MATERIALITY

Whale beachings are material events. Ocean currents and wave action contribute to an ever-changing shoreline. That which remains deposited on a shoreline during a receding tide each day or night offers beachcombers a treasure trove of interesting artifacts to explore and interpret. At low tide, beaches reveal marine life that ranges from vibrantly interactive tide pools teeming with life to dead, dying and inert substances. Moon cycles influence the cyclical action of the high and low tides. Thus, bio-debris along a shoreline is a normal state of affairs, particularly with a rich and abundantly diverse biosphere in the adjacent marine waters. Yet decomposing matter releases a pungent and sometimes foul odor, particularly when a large amount of biomass is involved. Small, decomposing marine mammals, such as seals, are commonly ignored or readily removed on public beaches by authorities who manage these public spaces. When a whale beaches, the massive body of these marine mammals presents an entirely new set of problems from a public affairs perspective.

The materiality of a beached whale cannot be ignored. Thousands of pounds of rotting biomass not only impedes the mobility of human traffic on a shoreline, it presents major removal problems due to the enormous size of the dead or dying whale. Furthermore, witnessing the stressful physiological situation of a beached whale can be emotionally taxing for humans who are attempting to help mitigate the stressful corporeal (and perhaps emotionally distressing) experience of the whale. The very subjectivity of a beached whale is disquieting even for those who might prefer to ignore it. For these reasons, and others described below, beached whales are major interruptions in daily beach life. They present material, symbolic and emotional challenges for various publics, particularly those in close physical proximity to beaches.

REPRESENTING WHALES

Whales have captivated humans for centuries. For example, Herman Melville's *Moby Dick*, a novel about an aggressive whale, was written in 1851 and is still considered a classic of American literature. Friendlier contemporary representations of whales also abound. For example, "Shamu" is the brand name for orca whales that entertain humans at Sea World, a land-based marine amusement park. And there are environmental campaigns to "Save the Whales." For example, *Whale Wars*, a reality series on the television channel Animal Planet, features Paul Watson and his team of activists on a ship called the *Sea Shepherd* as they fight with Japanese "research" ships to prevent the killing of whales in Antarctic seas.[2] In fact, whales are among a number of large magnetic mammals that science and technologies professor Benjamin Sovacool asserts are "charismatic mega-fauna," such as the polar bears and dolphins that are often featured in environmental campaigns at the expense of other critically important species within ecosystems (356). In short, whales have fascinated humans for centuries and remain subjects of human representation, entertainment and curiosity, as Tema Millstein explicates through her whale-watching field research.

Second, whale beachings have been recorded over the past several hundred years, but have become increasingly common since 1998 (Oceana; Swinford). These recent increases in whale beachings have caught the attention of activist environmental nonprofit organizations that describe these events in narratives competing with those of governmental and military institutions.

I have located hegemonic narratives in traditional print media articles from, for example, the *Times* (London), the *Los Angeles Times* and the *Washington Post* and compared them with counterhegemonic (nontraditional or alternative) new media website representations of whale beachings engaging whale activism (e.g., on the websites earthportals.com and nrdc. org). Through textual analysis, I have discerned two distinctly different narratives, and in this paper I provide exemplars of both.[3]

In the next section I analyze the deployment of beached whales as a rhetorical practice in two competing narratives (alternative new media and mainstream print media). Specifically, upon foregrounding the materiality of beached whales, I will explain with respect to each narrative (1) how the articulation of extraordinary materiality serves as a flashpoint for deliberative rhetorics; (2) the counterhegemonic and hegemonic discourses that comprise them (e.g., environmental and industrial/military); and (3) their (dis)connections with dialogic forms of "listening" to the varied ways in which the natural world speaks. I will begin with a narrative that deploys counterhegemonic discourses disseminated primarily through alternative, online media.

ALTERNATIVE NEW MEDIA NARRATIVES

Narratives found in alternative media configure the bodies of beached whales as tragic, unpleasant and material alarm systems that call attention to the accelerating degradation of ocean life and planetary life systems. Within these narratives, the bodies of beached whales are framed as both interlocutors and victims of greedy human perpetrators (i.e., those who are guilty of overconsumption of food sources through commercial fishing practices). And because whales cannot "speak" human, they require saving. From these loci, the beached whales' bodies articulate with human narratives that resonate with critical capitalist discourses and progressive ideologies calling for sustainable practices in the production and distribution of human goods. Willard Van de Bogart, a publisher for *Insights*, a "Conscious E-Zine," provides one example of the beached whale framed as both interlocutor and victim:

> The information you are about to read in this article is not pleasant, and is very sad. It questions the wisdom of the scientific quest, and makes you want to stop this unbelievable madness. . . . The internet has become a global voice and by presenting this awareness we can better question our own actions, and how that may impact our surro[u]ndings and our planet. It is clear beyond doubt that current ideology is allowing not only the killing of oc[ea]n life, but indigenous people, forests, rivers and waterways, and ultimately our own species. As writer and publisher for "Insights" I too try to express my opinions, as we all should, now that we have access to a global communications medium. If you think that our planet is not in serious trouble then this article will pin point a devastating look at what technol[o]gy and poor thinking is doing to the oldest living creatures on Earth. (n. pag.)

This is Van de Bogart's introduction to the essay by Bobbie Sandoz featured on the *Insights* website. Certainly, in this text, we can see that Van de Bogart foregrounds the significance of understanding human relationships with the natural world. He frames science and technology in a negative light and yet he values the Internet as a communication medium for global human voices. Moreover, he presumes that the "oldest living creatures" and the natural world require humans to save them from other human activities. Van de Bogart invokes alternative and counterhegemonic environmental discourses that privilege sustaining healthy natural systems and a rich diversity of life-forms on the planet in opposition to hegemonic capitalist discourses that privilege efficiency, productivity and profits. Van de

Bogart, like many environmentalists, is skeptical about human capabilities to "listen" to the natural world and hear the varied ways in which natural systems and extrahuman Others express themselves (Arnett). Also in this narrative, we can see that the "oldest living creatures" (or whales) that he references function as a flashpoint for the deployment of a rhetorical appeal to move humans to speak out and act on behalf of the whales and a planet "in serious trouble." Let's now turn to an analysis of Bobbie Sandoz's story, which Van de Bogart introduces.

Sandoz takes up this topic of paying attention to beached whales as a form of listening to the natural world. Moreover, she uses a metaphor as a rhetorical appeal to her Internet audience, suggesting that beached whales might be environmental warning signals like canaries in coal mines.[4] This invokes ecocentric discourses that are counterhegemonic, because they oppose human dominance over nature and position whales as natural subjects with inherent value and communicative capabilities right alongside humans. In the narrative that follows, Sandoz juxtaposes distressed and dying cetaceans with healthy cetaceans, characterizing the latter as interlocutors capable of developing friendships with noncetaceans (in this case, humans). Ironically, in order to posit these transcetacean relational and communicative capabilities, Sandoz imbues orca whales with anthropomorphic qualities such as the ability to forgive, play and perform acts of kindness. She writes:

> In January, hundreds of dolphins stranded along our eastern coast, while a superpod of orcas off California circled a group of boats to flirt and play with the people aboard, then remained afloat on the surface as the sun set on their fins. A similar contrast occurred last summer when a group of San Juan Orcas appeared in time to appreciate a human concert given in their honor, showing particular interest in Amazing Grace, or forgiveness without merit, while hundreds of their gray whale cousins washed up dead on west coast shores. . . . To add to the disquiet of these events, independent scientists strongly disagree with government-funded "experts" on the true effect of an extremely intense and apparently lethal force going into the ocean called low frequency active sonar (LFAS) that appears to be leaving a trail of corpses in its path. . . . Will we remain cooperatively in denial, uninformed, and dangerously silent in the face of this potential planetary holocaust as we have in the past, or will we awaken in time to stop this serious threat to all marine life? And, if we fail to respond, what will the consequences be for our own survival and souls? (n. pag.)

In Sandoz's narrative we can see that the material articulation of "hundreds of gray whale" corpses on beaches (framed as "cousins" juxtaposed with a healthy and happy "superpod of orcas") serves as a flashpoint or node whereby Sandoz deploys a rhetorical appeal to her (human) Internet audience. Her appeal exercises *logos, ethos* and *pathos* and it targets human denial and complacency in the face of a "serious threat to all marine life"— that of low frequency active sonar (LFAS), a "lethal force going into the ocean." The reference to LFAS does not explicitly name the U.S. Navy as the antagonist in this story, but this allegation is tacit, because the narrative was written in the context of an environmental controversy over low- and mid-level active sonar naval testing practices. Thus, this rhetorical appeal invokes postcolonial discourses that critique the hegemonic practices of the U.S. military-industrial complex. Sandoz implies that these practices are staging nothing less than a "potential planetary holocaust" with implications for human survival both physically and spiritually (n. pag.).

Herein, we can see that whale rhetoric, deployed from the corpses of hundreds of beached gray whales, mobilizes discourses that are critical of governmental apparati that prop up neocolonial practices in the global production and distribution of goods.[5] These practices, Van de Bogart and Sandoz both claim, are not only marginalizing and harming subaltern and indigenous peoples, but also degrading planetary systems upon which humans ultimately depend.

In sum, in these counterhegemonic narratives, the beached whale bodies are described by empathic rhetors who narrate a critique of human relationships with the natural world. This critique deploys a rhetorical call to action that connects with discourses of environmental sustainability to heighten awareness of a planet in jeopardy (albeit with an anthropocentric spin). For example, Van de Bogart contends that public participation is imperative to "stop the unbelievable madness." Thus, deliberative rhetorics deployed from the bodies of beached whales connect with counterhegemonic discursive formations that antagonize hegemonic ideologies. Tracing these trajectories demonstrates the material rhetoricity of beached whales. In Sandoz's narrative, stranded cetaceans are a form of "voice" that demands our attention. How humans come to understand such extraordinary material articulations is deeply connected to the ways in which we (do not choose to) *listen* to natural voices. It is connected as well to the rhetorical deliberations that follow from the collision of narratives that assume unique forms of listening to natural voices (and those that do not).[6] In the next section, I will examine hegemonic narratives of beached whales in contrast to the latter narratives, as represented in the mainstream print media.

MAINSTREAM PRINT MEDIA NARRATIVES

There is relatively little representation of beached whales in the mainstream national U.S. print media. Generally, stories emerge in national print media in proximity to the event of a beaching. Print articles reveal that mass beachings in 2007 and 2008 occurred on shorelines adjacent to naval sonar testing areas. The dominant media narrative for beached whales represents them as an anomaly and renders the bodies of beached whales as scientific specimens comprised of inert tissue for analysis. As such, scientific authorities are called upon to explain the causality of these displaced and objectified bodies. Unlike the alternative media narrative, which invokes compassion for the victimized whale as a subject, the dominant media narrative configures these massive displaced bodies as problematic, a nuisance or an interruption, in part because of their sheer size and removal costs, but more importantly because their increased occurrences resulted in a court injunction that literally interrupted military sonar testing exercises. In what might be characterized as an effort to distract from the emerging counterhegemonic discourses described above, voices of governmental and scientific authority are called upon to dissociate the whales' bodies from sonar testing and to attribute these scarred and bleeding carcasses to other causes. The dominant beached whale narrative configures the primary causes of whale beachings as the natural consequence of unknown diseases, navigational problems and accidental collisions with ships.

For example, an article in the *New York Times* reports, "Mass strandings are especially common among pilot whales. . . . Scientists theorize that the animals may lose their sense of navigation while feeding or follow a sick animal that has gone astray" ("Rescuers Lead" A16). Another *New York Times* article completely erases the subjectivity of a finback whale that ended up on some rocks across from President Bush's vacation home. It states, "The carcass of a finback whale . . . has defied attempts to dynamite it into pieces to keep it from returning to shore. The explosives did little more on Friday than send up smoke and put holes in the whale. . . . A police dispatcher said the whale was still floating today" ("Carcass Of" A15). The objectification of whale bodies in hegemonic narratives enables scapegoating or the shifting of blame away from anthropogenic causes of death toward natural or organic ones. It also makes it easier to treat the bodies of whales as inert substances with little empathy for their subjectivity and corporeal experience of dying. The articulations of whales on shorelines encountered by humans that carry dominant ideologies deploy rhetorics of inconvenience that get mapped onto discourses of national security and military action (such as collateral damage).

Put another way, in the narratives that I offer below, stranded and injured whales become a flashpoint for the deployment of whale rhetorics that connect with hegemonic discourses. First, dealing with these enormous bodies becomes an ordeal in itself, and those working to prevent the public from having to cope with the stench from rotting carcasses are characterized as heroes. In contrast to nongovernmental scientific findings that link whale beachings to naval sonar testing, the hegemonic narrative eschews these findings until it is revealed (through deliberative rhetorics) in a court case, filed by the National Resource Defence Council (NRDC) against the U.S. Navy, that naval sonar testing is associated with mass beachings. At this juncture, the bodies of whales become necessary collateral damage or the unfortunate result of friendly fire in the war on terror. Moreover, as Marc Kaufman of the *Washington Post* informs us, the articulation of whales as collateral damage includes the voice of former commander in chief George W. Bush, who, according to governmental defense lawyers in the aforementioned court case, determined that "allowing the use of mid-frequency sonar in ongoing exercises off Southern California was 'essential to national security' and of 'paramount interest to the United States'" (A02).

Los Angeles Times staff writer Steve Chawkins captures the essence of the beached whales as problematic interruption narrative—and objectified scientific specimen—in his coverage of a series of recent blue whale beachings in Southern California. He describes scientists' efforts to speed up the process by towing the whale's body to a beach at Naval Base Ventura County, "enabling them to scrutinize the whale's tissues before they further decompose" (Chawkins 3). He mentions that the site "was chosen so the public would not be exposed to the odor or the expense of getting rid of it."

> Hauling the carcass through the choppy channel was expected to take up to 10 hours. From the air, the vessels—perhaps 30 feet long—looked like dinghies next to the carcass, which was scraped up and bleeding. The boats took up positions on both sides, and crew members played out a looped sling and pulled it tight around the tail. One of the vessels would haul the load, weighing 50 tons or more, while the other accompanied it. Scientists today plan to start a detailed necropsy. The process could last through the weekend and is crucial if biologists are to determine what is killing the whales. (Chawkins 3)

In this excerpt, we can see that the whale's body is represented as "tissue" and characterized as a bloody inert object. Measures are being taken to prevent the fifty-ton "carcass" from beaching, so as to enable a "detailed necropsy," a medical procedure to determine causality. Another excerpt from Chawkins's article further demonstrates the dominant narrative in which the whale's body is configured as object, dead and inert, due to natural or accidental causes:

> On Friday, the National Marine Fisheries Service urged vessels plying the channel to report dead whales they have seen or inadvertently hit. Even with blue whales weighing 60 tons or more, massive freighters and tankers can run into them without realizing it.
> "On a vessel doing 40 knots, no one notices if they've dropped half a knot," Mate said. "Sometimes it's only when they get into port that a huge, rotten mass floats off." (Chawkins 3)

In both of these examples there is evidence of the desire to distance these massive bodies from the public eye and public spaces. The potential displacement of these objectified marine bodies onto beaches is framed as problematic and costly due to the massive corporeality of the "carcass." Thus, preventing a dead or dying whale from beaching avoids a material rupture (which serves as a flashpoint in the deployment of counterhegemonic discourses) and the arduous removal of the decomposing body from the adjacent shore.

Unlike the counterhegemonic narratives, the hegemonic narratives leave little room for apprehending the whale as a subject capable of voice. As such, any notion of "listening" to beached whales as material rhetorical articulations is absent. Instead, the bodies are instantly hailed as specimens to be analyzed through scientific and objective endeavors that begin with hypotheses that do not include sonar as a plausible cause.

The next two examples demonstrate a more recent corollary to this dominant narrative—beached whales as 'necessary collateral damage.' These examples relate to the growing scientific evidence that mass strandings of whales suffering from decompression sickness (from surfacing too fast—a condition that also affects human divers) are linked to military sonar exercises in adjacent waters (Jepson et al.). In the *Washington Post,* Marc Kaufman writes:

> The government filings said the federal ruling limiting sonar use "profoundly interferes with the Navy's global management of U.S. strategic forces, its ability to conduct warfare operations, and ultimately places the lives of American sailors and Marines at risk."
>
> "We cannot in good conscience send American men and women into potential trouble spots without adequate training to defend themselves," [Adm.] Roughead said. "The southern California operating area provides unique training opportunities that are vital to preparing our forces, and the planned exercises cannot be postponed without impacting national security." (Kaufman A08)

Joe Mozingo's article, appearing in the *Los Angeles Times* in 2007, also configures beached whales as necessary collateral damage:

> "The ability to detect and track potentially hostile submarines is a critical skill that cannot be duplicated in the classroom or by simulation," said Navy spokesman Capt. Scott Gureck. . . . The Navy says more than 40 nations, including Iran and North Korea, use quiet diesel submarines that are best detected using bursts of mid-frequency sonar. . . . Adm. Robert Willard, commander of the U.S. Pacific Fleet, said in a statement, "The initial injunction left us in an untenable position of having strike groups needing this training and not being able to accomplish it. (Mozingo 1)

The latter two examples clearly frame the beached whales as material interruptions in mid-level frequency naval sonar exercises that are necessary for national security. Admiral Willard's remark clarifies that the 2007 judicial injunction to temporarily stop mid-level frequency sonar testing impedes the training of naval and marine strike groups. Thus, dominant narratives of beached whales as inert tissue for analysis deploy scientific analysis discourses, as well as patriarchal discourses that prop up U.S. homeland security practices and neocolonial (imperial) nation-state ideologies.

Both Iran and North Korea are invoked in the latter argument, where beached whales as empty signifiers articulate as collateral damage in President Bush's war on terror. Both Iran and North Korea are potentially positioned in U.S. crosshairs, because they are both located in what the Bush administration refers to as the "axis of evil." For example, the Iraq War, in which tens of thousands of people have died and been displaced, followed on the heels of identifying Iraq as a country within this so-called "axis."

In this contemporary age of terrorism and culture of fear, dominant discourses of homeland security seek to stabilize the nation-state and patriarchal discourses shore up safety and protection at the expense of acoustically sensitive marine mammals. The materiality of displaced whale bodies on beaches literally threatens the ability for the military to keep the homeland safe. The purported consequence of this recent judicial stay, which grants the military an exception to the Endangered Species Act (to continue these potentially harmful sonar practices), is the promise of a safe and secure America. The consequences from the deployment of these discourses, of course, include harmful material realities for whales, Iraqis (and potentially Iranians and North Koreans), human communities in the area where the naval sonar testing occurs and the marine biosphere.

Thus, we can comprehend Ronald Greene's notion of a materialist rhetoric, which "offers a way to produce a cartography of deliberative rhetoric without reducing its effectivity to the politics of representation" (22). We can identify nodes wherein rhetoric becomes a discourse of power. Disparate beached whale rhetorics deployed from the loci of extraordinary material articulations (e.g., whales as canaries or whales as collateral damage) mobilize a matrix of (counter)hegemonic discourses on a field of material and discursive structures (e.g., in media and in governmental institutions, in this case a court of law) that host and sponsor rhetorical deliberations. The outcomes of these deliberative rhetorics manifest in material consequences that get written back onto whale bodies (e.g., beachings due to physiological manifestations from surfacing too fast) as well as onto other acoustically sensitive cetaceans in the marine biosphere.

This dynamic way of apprehending beached whales does not lose track of the materiality of articulations. Greene's theoretical framework helps us conceptualize materially imbricated articulations that enable and constrain certain discourses, ideologies, social action and the interplay among them. This matrix includes real and/or purported consequences for whales, military practices and possibly human populations in nations framed as "evil" as well as for living systems within which we are all embedded. In short, we can analyze how the dynamic interplay of articulations that deploy rhetorics connected to (counter)hegemonic discourses is simultaneously deployed by and accordingly mapped back onto beached whales. In the case of counterhegemonic narratives, we have seen that whales are hailed as friendly subjects and imbued with anthropomorphic characteristics. In the case of hegemonic narratives, whales are framed as disruptive inert objects or as collateral damage on beaches near sonar training practices. Each of these represents two very different ideologies and human relationships with nature.

IMPLICATIONS

Mapping a cartography of rhetorical deployments from the bodies of beached whales can uncover distinctly different ideologies of human relationships with the natural world. But, in doing this, what additional tensions arise?

Admittedly, this analysis remains tethered to (inexorably human) texts, which must be fixed or held frozen in time, at least temporarily, in order to make possible an analysis or identification of articulations to beached whale bodies, from which rhetorics are deployed and discourses are mobilized. I have demonstrated how foregrounding extraordinary material articulations can shift the focus of rhetorical criticism in a direction where it can begin to account for animals and the natural world. Doing this, though, does risk reinscribing a similar objectification on the material articulations that are selected for analysis. In other words, my analysis of breached whales as windows into understanding two different ideologies about animals and nature robs these whales of their subjectivities. Analysis, by definition, however, involves a certain degree of objectification and distance. This paradox is difficult to escape and certainly some tensions remain. There are additional considerations.

Human sonar technologies are derived from studying cetacean communication. This has implication for rhetorical critics interested in postcolonial scholarship. In other words, could there be a more vivid example of colonization? Moreover, dominant discourses reassure us that homeland security trumps the Endangered Species Act. On the other hand, if we think of hegemony as participation in one's own oppression, who among us does not want to feel safe? Are beached whales, then, articulations of collateral damage, or are they radical and unruly bodies/arguments that provide an opportunity to mobilize discourses that assert, "I have met the enemy and we are it"?

It remains to be seen how the matrix of competing alternative and dominant narratives will play out, over space and time, in shaping human relationships with whales, the natural world and alien Others. Through this analysis, however, we can reassert that the material bodies of beached whales impose themselves upon us and call forth rhetorical responses. Given the different responses in the (counter)hegemonic discourses that I have analyzed, the question remains: What are some other possibilities for how we can relate to and understand beached whales? I agree with Rogers that a dialogic response might be helpful, because it suspends the impulse to judge and blame. Rather, it seeks to genuinely understand radically different perspectives through listening.

As I have alluded to throughout this analysis and as other environmental communication scholars are beginning to address, honing our skills as *deep listeners* to the natural world might be another possibility. In dialogues, within which I have participated and facilitated, a kind of deep listening occurs that makes space for imagining radically disparate realities. I know that I often emerge from dialogic moments transformed, spoken anew, through listening to Others. What might we stand to gain if dominant practices valued the voices of nature, and what might we stand to gain if we found ways to learn from those who *practice* listening to nature? This moves us in the direction of transhuman dialogue, as Richard Rogers suggests. Put another way, what can we hear and learn from analyzing more extraordinary material articulations?

CONCLUSION

Greene's materialist rhetoric combined with analysis of media representations enables a fluid and dynamic critical analysis of the deployment of beached whale rhetorics onto a field of discourses within a broad context. It offers a way to foreground the dynamism of beached whales' bodies intertwined with human communication and socially constructed realities that have implications for how rhetorical theory and criticism can begin to deal with and account for animals and nature.

As described above, this milieu of social discourses can be mapped and traced back to the extratextual realities (physical bodies) from whence they came and far beyond into the marine biosphere and other nations. Furthermore, by looking at both hegemonic and counterhegemonic narratives (in both mainstream and alternative media) that emerge from beached whales, we can begin to understand the mutually constitutive aspects of material and discursive formations and the ways in which they get ignored or privileged. In choosing extraordinary material articulations for analysis we can also gain insights into diverse human relationships with the natural world. This risks further objectification of extrahuman Others, but it can also create openings and illuminate pathways toward more dialogic relationships.

Whale beachings are material ruptures, openings for rhetorical invention by speech communities. Turning an eye toward and lending a carefully attuned ear to the rhetorical reverberations of extraordinary and extrahuman material articulations may be one way for us to move in the direction of a transhuman dialogue. Moreover, tracing a cartography of deliberative rhetorics from extraordinary and extrahuman material articulations is another way for rhetorical criticism scholars to begin dealing with nature and the natural systems in which we are all embedded. At this juncture in history, commensurate with climate change, depleted supplies of potable water and the ongoing radical reduction of biodiversity on the planet, finding practical ways to deconstruct environmental exigencies and develop more holistic and sustainable practices is imperative.

NOTES

1. See, for example, Anna Lowenhaupt Tsing's ethnography *Friction: An Ethnography of Global Connection* (238).
2. The reason "research" appears in quotes here is that a 1986 International Whaling Commission moratorium prevents the killing of whales for commercial purposes. The *Sea Shepherd* confronts the Japanese research vessels in Antarctic waters purporting that whale meat continues to be sold in Japanese commercial fish markets under the guise of research. Multiple international lawsuits exist (see "Australia Heads to Court")
3. Admittedly, analysis of media texts reifies the foregrounding of discursive representations. Future research could include participant observation at the site of whale beachings, where whale/human elemental encounters preexist articulation to mediated rhetorics (for examples of such participatory research, see Blair; Endres, Peterson, and Sprain; and Pezzullo).
4. When a canary dies in a coal mine, miners know that they need to get to a location with better air quality. This human 'technology' keeps humans safe at the expense of canaries. As such, the dead or physically compromised canary serves as a warning that the physical environment is not conducive for sustaining life.
5. For cross-disciplinary approaches to capitalist criticism see, for example, Deetz; Giddens; and David Harvey.
6. For more information on unique forms of listening, see the works of environmental communication scholars Donal Carbaugh ("'Just Listen'" and "Quoting the Environment"); Michael Salvador; and Julie Schutten and Richard Rogers.

REFERENCES

Arnett, Ronald. "A Dialogic Ethic 'between' Buber and Levinas: A Responsive Ethical 'I.'" *Dialogue: Theorizing Difference in Communication Studies.* Ed. Rob Anderson, Leslie Baxter, and Kenneth Cissna. Thousand Oaks, CA: Sage Publications, Inc., 2004. 75–91. Print.

"Australia Heads to Court over Whaling Dispute with Japan." *CNN* 1 Jun. 2010. Web. 1 Mar. 2012.

Biesecker, Barbara. "Michel Foucault and the Question of Rhetoric." *Philosophy and Rhetoric* 25.4 (1992): 351–364. Print.

Blair, Carole. "Reflections on Criticism, and Bodies: Parables from Public Places." *Western Journal of Communication* 65 (2001): 271–294. Print.

Carbaugh, Donal. "'Just Listen': 'Listening' and the Landscape among the Blackfeet." *Western Journal of Communication* 63.3 (1999): 250–270. Print.

————. "Quoting the Environment: Touchstones on Earth." *Environmental Communication: A Journal of Nature and Culture* 1.1 (2007): 64–73. Print.

"Carcass of Beached Whale Survives Dynamite Blast." *New York Times*. 26 August, 1991, sec. A.

Chawkins, Steve. "Blue Whale Is Towed to Navy Base; to Preserve Evidence, Scientists Want to Quickly Begin a Necropsy That They Hope Will Tell Them Why Three Such Mammals Have Died." *Los Angeles Times* 22 Sept. 2007: CA Metro, B3. *LexisNexis Academic*. Web. 24 Feb. 2008.

Cloud, Dana. "The Materiality of Discourse as Oxymoron: A Challenge to Critical Rhetoric." *Western Journal of Communication* 58 (1994): 141–163. Print.

Cox, Robert. "Nature's 'Crisis Disciplines': Does Environmental Communication Have an Ethical Duty?" *Environmental Communication: A Journal of Nature and Culture* 1.1 (2007): 5–20. Print.

Deetz, Stanley. *Democracy in an Age of Corporate Colonization: Developments in Communication and the Politics of Everyday Life*. Albany: SUNY P. 1992. Print.

DeLuca, Kevin. *Image Politics: The New Rhetoric of Environmental Activism*. New York: The Guilford Press, 1999. Print.

————. "Unruly Arguments: The Body Rhetoric of Earth First!" *Argumentation & Advocacy* 36.1 (1999): 9–21. Print.

Egelko, Bob. "Supreme Court on Sonar: Navy Trumps Whales." *San Francisco Chronicle* 13 Nov. 2008. Web. 24 Feb. 2012.

Endres, Danielle, Tarla Rai Peterson, and Leah Sprain, eds. *Social Movement to Address Climate Change: Local Steps for Global Action*. New York: Cambria Press, 2009. Print.

Foucault, Michel. *Power*. Ed. James D. Faubion. Vol. 3. New York: The New York Press, 1994. Print.

Giddens, Anthony. *Capitalism & Modern Social Theory: An Analysis of the Writings of Marx, Durkheim and Max Weber*. Cambridge: Cambridge UP, 1971. Print.

Greene, Ronald. "Another Materialist Rhetoric." *Critical Studies in Mass Communication* 15.1 (1998): 21–40. Print.

Hall, Stuart. *Representation: Cultural Representation and Signifying Practices*. London: Sage Publications, 1997. Print.

Harvey, David. "The Right to the City." *International Journal of Urban and Regional Research* 27.4 (2003): 939–941. *Wiley Online Library*. Web. 1 Mar. 2012.

Jepson, P. D., et al. "Gas-Bubble Lesions in Stranded Cetaceans: Was Sonar Responsible for a Spate of Whale Deaths after an Atlantic Military Exercise?" *Nature* 425 (2003): 575–576. Web. 1 Mar. 2012.

Kaufman, Marc. "Navy Wins Exemption from Bush to Continue Sonar Exercises in Calif.; President Cites National Security in Order." *Washington Post* 17 Jan. 2008, correction appended suburban ed.: A02. *LexisNexis Academic*. Web. 24 Feb. 2008.

McGee, Michael Calvin. "The 'Ideograph': A Link between Rhetoric and Ideology." *Quarterly Journal of Speech* 66 (1980): 1–16. Print.

McKerrow, Raymie. "Critical Rhetoric: Theory and Praxis." *Communication Monographs* 56.2 (1989): 91–111. Print.

Milstein, Tema. "When Whales 'Speak for Themselves': Communication as a Mediating Force in Wildlife Tourism." *Environmental Communication: A Journal of Nature and Culture* 2.2 (2008): 173–192. Print.

Mozingo, Joe. "Navy Can Use Sonar, Court Says; Appellate Ruling Gives the Go-ahead for Training off the Southland Coast, Despite Concerns That Whales

May Be Harmed." *Los Angeles Times* 1 Sept. 2007: CA Metro, B1. *LexisNexis Academic*. Web. 24 Feb. 2008.

Norton, Bryan. "Environmental Ethics and Weak Anthropocentrism." *Environmental Ethics* 6.2 (1984) 131–148. *University of Utah Philosophy Documentation Center Collection*. Web. 5 Jul. 2012.

Oceana. "The Death of Cetaceans through the Use of LFA Sonar in Naval Military Maneuvers." Aug. 2004. Web. 23 Feb. 2008. <http://www.oceana.org/fileadmin/oceana/uploads/europe/reports/stranding_of_cetaceans.pdf>.

Pezzullo, Phaedra. "Resisting National Breast Cancer Awareness Month: The Rhetoric of Counterpublics and their Cultural Performance." *Quarterly Journal of Speech* 89 (2003): 345–365. *Ebscohost Communication & Mass Media Complete*. Web. 5 Jul. 2012.

Rogers, Richard. "Overcoming the Objectification of Nature in Constitutive Theories: Toward a Transhuman, Materialist Theory of Communication." *Western Journal of Communication* 62 (1998): 244–272. *Ebscohost Communication & Mass Media Complete*. Web. 5 Jul. 2012.

Salvador, Michael, and Tracylee Clarke. "The Weyekin Principle: Toward an Embodied Critical Rhetoric." *Environmental Communication: A Journal of Nature and Culture.* 5.3 (2011): 243–260. Print.

Sandoz, Bobbie. "Beware the Beached Canaries: What Stranded Dolphins and Whales Are Trying to Tell Us about a Global Emergency beneath the Sea." *Insights*. Web. 23 Feb. 2008. <www.earthportals.com/beachedwhales.html>.

Schutten, Julie, and Richard Rogers. "Magick as an Alternative Symbolic: Enacting Transhuman Dialogs." *Environmental Communication: A Journal of Nature and Culture* 5.3 (2011): 261–280. Print.

Simons, Herbert, ed. *The Rhetorical Turn: Invention and Persuasion in the Conduct of Inquiry*. London: U of Chicago P, 1990. Print.

Sovacool, Benjamin. "Spheres of Argument Concerning Oil Exploration in the Arctic National Wildlife Refuge: A Crisis of Environmental Rhetoric?" *Environmental Communication: A Journal of Nature and Culture* 2.3 (2008): 340–361. Print.

Swinford, Steven. "Navy Adapts Sonar to Protect Whales." *Sunday Times* (London) 26 Mar. 2006, sec. home news: 7. *LexisNexis Academic*. Web. 23 Feb. 2008.

Tsing, Anna Lowenhaupt. *Friction: An Ethnography of Global Connection*. Princeton: Princeton UP, 2005. Print.

Van de Bogart, Willard. "The Killing of Earth's Whales by Sonar: US Navy, and NATO Test Low Frequency Active Sonar LFAS System to Detect Diesel and Nuclear Submarines." *Insights*. Web. 23 Feb. 2008. <www.earthportals.com/beachedwhales.html>.

Wander, Phillip. "The Ideological Turn in Modern Criticism." *Central States Speech Journal* 34.1 (1983): 1–18. Print.

———. "The Third Persona: An Ideological Turn in Rhetorical Theory." *Central States Speech Journal* 35.4 (1984): 197–216. Print.

Zezima, Katherine. "Rescuers Lead 46 Beached Whales Back to Deeper Waters," *New York Times*. 30 July 2002, sec. A.

4 Framing Primate Testing
How Supporters and Opponents Construct Meaning and Shape the Debate

Joseph Abisaid

INTRODUCTION

Historically speaking, animal researchers have enjoyed broad public support in their work. A recent Gallup poll of Americans found that almost 6 in 10 respondents agreed that medical testing on animals was morally acceptable ("Gallup Poll"). In countries that have traditionally been less accepting of animal experimentation, similar findings have been reported (Williams). Yet, true public opinion on animal experimentation remains difficult to gauge because of the different species used in experiments, the researchers' goals, and the reported conditions of the animals being experimented on (Goldberg 545). This point is best illustrated by other polls that have been conducted that show wavering support for animal experimentation. For example, a recent Pew Research Center Poll found that only 52% of the general public is in favor of using animals in scientific research ("Public Praises Science") and according to another recent poll commissioned by Physicians Committee For Responsible Medicine, there is overwhelmingly strong opposition for testing cosmetic products on animals ("More Than A Makeup Trend"). These survey findings suggest two important points. First, public opinion on animal experimentation is malleable and subject to change with the right sets of arguments. Second, how the arguments for and against animal testing are framed by supporters and opponents may influence how individuals think about animal experimentation. Therefore, not only is there a deep division between supporters and opponents on the use of animals in scientific research, but there also exists a strong belief by both sides that they can influence the general public to support their cause.

The use of nonhuman primates for experimental purposes (hereafter referred to as primate testing) represents a very interesting case study for how supporters and opponents frame the debate about animal research. Primate testing refers to the experimental use of primates via invasive and uninvasive means of research inquiry, but almost always for the benefit of humans (LaFollette and Shanks). While all animal-based research raises serious ethical and moral questions, primate testing puts individual and societal ethics and morals to the test because of the strong evolutionary and genetic similarities between primates and humans as well as the uncontroversial fact that primates possess advanced cognitive abilities, experience a range of feelings similar to humans and form social relationships that contain elements of compassion, altruism, empathy and love.

Many studies have demonstrated these similarities. For example, a study by a group of decision-making theorists and psychologists found that primates possess advanced cognitive abilities that include making rational decisions, experiencing cognitive dissonance and thinking their way out of it (Egan, Bloom, and Santos 204). A study by two neuroscientists at Duke University demonstrated that monkeys can mentally add the numerical values of two sets of objects and select a visual array that is approximate to the arithmetic sum of the two sets. In addition, monkeys exhibited the same pattern of selection as humans who were tested on the same nonverbal tasks. The researchers concluded that nonverbal arithmetic is not exclusive to humans but rather part of an evolutionary primitive system for mathematical thinking that is shared by humans and monkeys (Cantlon and Brannon). Another experimental study by Joshua Redford, a cognitive psychologist, found that monkeys engage in metacognitive monitoring. That is, they have the capacity to monitor their cognitive processes. The findings showed that both humans and monkeys increased their cognitive effort during high-difficulty categories, suggesting that both have the ability to control their metacognitive processing.

Other studies have shown primates to be similar to humans in how they use tools. For example, a group of psychologists and biologists working in Thailand observed that monkeys were skilled tool users and used stone tools regularly (Gumert, Kluck, and Malaivijitnond). Monkeys consciously chose tools with varying qualities, including pounding hammers for crushing shellfish and nuts as well as anvils and axe hammers for oysters. Another study by a group of researchers at the University of Wisconsin found that, contrary to what was previously thought, monkeys were found to be self-aware when standing in front of a mirror (Rajala, Reininger, Lancaster, and Populin). Monkeys exhibited self-directed behaviors and demonstrated specific social responses. The researchers theorized that rather than a cognitive divide between humans and primates, there is more of an evolutionary continuum of mental functions. For these reasons and others, as Shirley Strum has pointed out, primate testing has become very controversial.

Despite these similarities, researchers insist that the experimental use of primates is important for building scientific knowledge. Concerning the role of chimpanzees specifically in biomedical research, VandeBerg and Zola write, "many advances from biomedical research with chimpanzees have been published in the past one or two years, demonstrating that rapid medical progress pertinent to a wide range of human diseases is being made through the use of chimpanzees" (32). Yet, research evidence points to the contrary.

A 2007 report by Bailey, Knight and Balcombe investigated the contribution of chimpanzee research to biomedical knowledge and its efficacy in combating human diseases. They identified 749 research articles between 1995 and 2004 that reported use of captive chimpanzees. From those studies, 95 were randomly selected and subjected to a detailed citation analysis. The researchers found that of the 95 research articles, 47 had yet to be cited in the scientific literature, 34 had been cited in 116 research papers that did not describe methods for combating human diseases, and the remaining 14 had been cited by 27 research papers that described methods for combating human diseases. The researchers then closely examined the 27 research papers to see what role primate testing had in combating human diseases. It was found that *in vitro* research, human clinical and epidemiological investigations, molecular assays and methods and genomic studies contributed to virtually all of the development. It was concluded that research studies involving chimpanzees—contrary to what advocates of primate testing maintain— played no meaningful role in combating human diseases. The researchers write regarding the results, "our review identified no studies of captive chimpanzees that made an essential contribution, or, in a large majority of cases, a significant contribution of any kind" (22). They conclude that primate testing has actually had an adverse effect on combating human disease: "far from augmenting biomedical research, chimpanzee experimentation appears to have been largely incidental, peripheral, confounding, irrelevant, unreliable and has consumed considerable research funding that would have been better targeted elsewhere" (23). A major scientific report by the Institute of Medicine in 2011 reached a similar conclusion that most current biomedical research involving chimpanzees is not necessary. In response to these findings, the National Institute of Health, a major source of primate-testing funding, issued a 2011 press release stating that "it will not fund any new projects for research involving chimpanzees while the Agency considers and issues policy implementing the IOM's recommendations." In short, primate testing is being seriously challenged in the public sphere.

The purpose of this chapter is not to settle the debate on primate testing, to decide whether it is right or wrong, fair or unjust, or scientifically beneficial or unnecessary, but rather to demonstrate how the groups involved on both sides of the debate offer dramatically different interpretations of the issue. These interpretations, in turn, lead to differing media frames presented to the public in hopes of swaying them to adopt a position. I analyze the framing of primate testing as it appears through various media channels promoted by organizations for and against primate testing.

Framing serves as the theoretical framework that guides my analysis of the primate-testing debate. A thorough examination of the message frames used in the primate-testing debate serves two important purposes for communication researchers. First, it helps them to identify the main themes, values, appeals, images, words and phrases that are invoked by both sides as they try to rally support. This is important because understanding the different frames helps us to understand the discourse surrounding the debate. Second, the analysis offers a way of understanding how humans think about their relationships with primates. That is, what does it mean to be human and what does it mean to be a primate— and how are the two interconnected?

I begin by briefly reviewing the literature on frames and framing. Especially important is the idea of contested frames, in which advocates from both sides disagree not only over what should be done about the issue, but over how people should think about it. I analyze frames that have been used in the primate-testing debate, relying on a select subset of frames that have appeared in various organizational materials put forth by those groups supporting and opposing primate testing. Specifically, I will look at specific pamphlets from both sides and offer interpretations of what they mean within a media framing context and what they mean in a broader context for the animal-human relationship. As will be shown, supporters and opponents frame primate testing very differently, with both groups claiming the moral high ground on the issue.

FRAMING AND THE CONSTRUCTION OF MEANING

The concept of framing enjoys widespread interest among scholars of communication theory (see, e.g., Entman; Pan and Kosicki). Researchers are in agreement with George Lakoff that framing constitutes an important and necessary part of successful communication, especially as it relates to highly contentious issues. Yet, as Dietram Scheufele argues, despite the central role of framing in the human communication process, researchers have struggled to fully understand it at both the conceptual and empirical levels. The absence of a unifying theory that guides framing research has led researchers to apply different conceptual and methodological approaches in their study of framing (D'Angelo). Since no agreed-upon definition or operationalization of frames or framing exists among researchers, it is important that framing studies offer a clear interpretation of how they conceptualize and investigate framing and frames.

Frames exist at two separate levels. First, frames can be thought of as mental cognitive representations inside an individual's mind. Pan and Kosicki refer to frames as "cognitive devices used in information encoding, interpreting, and retrieving" ("Framing Analysis" 57). Frames as mentally stored guides assist individuals in figuring out how they should understand and interpret their social environment. For example, a person with a strong animal rights frame makes sense of relevant issues as they relate to the treatment of animals rather than their property status or potential profitability. The other level of frames is characteristic of the media or news text. James Druckman refers to these types of frames as "frames in communication" or as attributes of the news or media itself. Within each news story is a frame or set of frames that serves not only to draw attention to the problem at hand but also to promote a particular way of thinking about and understanding the issue. The focus of the present study relies on the latter idea that frames in communication impact understanding of the issue through the process of selective control of information and the way in which messages are packaged to the audience. Media-level frames help to identify the problem, the social and political actors involved and the immediate solution. Thus, frames serve as the "central organizing idea or story line that provides meaning to an unfolding strip of events" (Gamson and Modigliani 143).

Through the use of words, phrases, images and symbols, both individuals and organizations give meaning to the issue. At the same time, the absence of certain concepts or terms within a media frame may serve to further enhance the importance of the frame in understanding the issue (Entman, "Framing: Toward"). In some instances, the omission of information is deliberate on the part of the framer. Other times, inconsistent information can be included in the frame as a way to address or refute alternative or competing frames. The constant reinforcement and repetition of certain well-placed frames make it more likely that media frames will be effective in developing a sense of understanding and meaning for individuals.

Part of the importance of being able to successfully frame an issue is rooted in the ability of frames to activate a particular train of thought. The resulting mental representations that occur upon exposure to a media frame are very critical in determining how successful that frame will be. Studies of news frames conclude that how we think about an issue is largely the result of how it is presented to us (Price, Tewksbury, and Powers; Valkenburg, Semetko, and de Vreese).

Besides influencing how individuals think about issues, frames have the power to shift public and political discourse and ultimately create social change. An ethnographic study by Valerie Jenness traced how the gay and lesbian movement successfully expanded the term "violence" to include violence toward gays and lesbians. By being able to control the discourse and define the terms of the debate, the movement has been able to redefine the issue and expand the domain of what is now classified as a "hate crime." Activists and organizations, including those working to eliminate primate testing, must surely take notice of the importance of shifting the discourse in a way that creates the most opportunity for social change.

As important as shifting the discourse is, being able to influence opinions and action through the manipulation of frames is even more important. Prior evidence shows that how controversial issues are framed can directly sway public opinion in one direction or another. For example, Entman, in his framing analysis of the KAL and Iran Air airline crashes, found that the frames dramatically differed: The Soviet downing of a Korean airliner was presented as deliberate and cold-blooded, while the U.S. downing of an Iranian plane was framed as a tragedy and a mistake resulting from miscommunication. Through an analysis of the words and images used in the news coverage, Entman concluded that the frames selected by the media not only serve to reinforce certain lines of thinking, but also have the capacity to impact public opinion ("Framing: U.S. Coverage").

Likewise, when news media present issues with certain status quo frames it can inhibit alternative frames, in turn leading to greater support for the dominant frame. For example, in an experimental study by Douglas McLeod and Benjamin Detenber that looked at protest news coverage of an anarchist rally, the authors found that the more coverage adhered to "status quo" frames, the more critical participants were of the protesters and the less likely they were to support their cause. News frame coverage that used alternative frames led to more support of the protesters.

Clearly, framing is an integral part of creating meaning and giving life to issues. Accordingly, then, frame construction and dissemination leads to increased frame-consistent thoughts, which in turn create shifts in public discourse that can ultimately influence social judgments and public opinion in the desired direction. For those advocating for or against primate testing, this serves as the basic framing model for persuasion. Within this particular study, only the first step is investigated. Future studies should consider the extent to which the frames communicated in the primate-testing debate shape public perceptions and other effect variables such as thoughts and attitudes. By focusing on the ways messages are framed in the primate-testing debate, I describe the discourse that surrounds the primate-testing controversy and how it configures the human-primate relationship.

Prior research suggests that the group best able to frame the issue will often be able to convince the public of the justness of their cause. As a consequence, framing should be considered as more than just a single aspect of the debate, but rather as the main focus for the social and political actors involved. Since no side has yet been able to wrest control of the issue, it stands to reason that the public either remains indifferent toward (or possibly just unaware of) the issue or that both sets of frames have been effective, therefore creating high levels of ambivalence among the public.

METHODOLOGY

The data were obtained from a public collection of promotional and organizational pamphlets put forth by advocacy groups for and against primate testing. Media frames showcasing support and opposition to primate testing were equally represented in the collection. Primate-testing support and opposition pamphlets were divided into two separate groups. Division of pamphlets was determined by affiliation of the organizational sponsorship and media content. In no instance was there any uncertainty regarding classification of materials. In order to determine which pamphlets from both sides would be used in the study, a randomization procedure was utilized for all available pamphlets in each group. Since many of the materials were fairly similar as well as repetitive in nature, it was decided to examine only a subset of the pamphlets.

It is difficult if not entirely impossible to categorize all the different frames in the primate-testing debate. To best account for the possible frames, I rely on the constructivist framework for "discovering" frames first set forth by William Gamson and Andre Modigliani (1987); a careful reading of relevant texts enables me to assign labels to different social categories (Pan and Kosicki, "Framing as a Strategic Action"). These different labels and meanings are derived from the unique framing devices that are part of the public discourse about primate testing (Pan and Kosicki, "Framing Analysis"). Texts, in this regard, refer to the pamphlets put forth by organizations working for and against primate testing.

MEDIA FRAMES IN SUPPORT OF PRIMATE TESTING

It is not surprising that there are several individuals and groups working to maintain the use of primates in experimental research. The Foundation for Biomedical Research cites primate testing as "extremely important" for many medical breakthroughs. Many researchers vigorously defend primate testing on the grounds that it has the ability to help cure diseases. More complicated and less publicized are the several individuals and organizations that are financially and professionally invested in primate research. So gigantic are the industries that promote primate testing—as well as animal testing in general—that cessation of such research would likely lead to hardship in the form of monetary loss or professional prestige since many primate researchers have received their core training through the experimental use of primates. Primate testing serves as a major source of revenue not just for researchers, but also for some universities and private research facilities. When researchers are able to secure grants for their research, the sponsoring institution takes a portion of that money regardless of whether the research produces any promising results. Clearly, there is good reason to continue with primate testing for those directly involved in it, as defenders of primate research have spent considerable economic resources to ensure that it continues (Greek and Greek).

Few frames, however, discuss the economic benefits that may result from primate testing. Rather, most frames focus exclusively on the scientific benefits of testing as well as the treatment of primates used for testing. A common theme of the frames that support primate testing is the idea that there is a direct relationship between primate-testing research and lives saved. Even though this notion itself has come under scrutiny among members of the scientific community and much of the research that is done with primates is referred to as basic research, which rarely translates into clinical utility success (Contopoulas-Ioannidis, Ntzani, and Ioannidis), this has not prevented such frames from frequently appearing. There is good reason to advance such a frame. Much of the public desires scientific progress, usually in the form of cures, drugs and longevity. The "scientific progress frame" plays to the audiences' hopes, fears and desires with the expectation that audience members will trust the connection between primate testing and scientific progress.

Figure 4.1 illustrates how groups in favor of primate testing employ frames that accentuate this relationship and that convey the positive treatment of animals used for testing. A brochure by the Foundation for Biomedical Research—the most influential of all animal research lobby groups—unequivocally states the relationship with the slogan "Animal research saves lives" centrally placed in big bold letters.

Included is a picture of what appears to be an attractive, young, healthy woman who is smiling and cuddling with a dog. The image itself is somewhat ambiguous. It is not clear who the young woman is. Is she a researcher

Figure 4.1   Foundation for Biomedical Research
brochure.

who conducts tests on animals or is she a recipient of medical treatment that was made possible through animal research? The dog in the picture also presents a kind of paradox for the reader. Dogs occupy a special place for many humans and this bond is made clear through the image, yet dogs are also subjected to harmful experiments. There are several different ways to interpret this image. One possibility is that the young woman is vibrant and healthy looking, a direct result of animal research. Another possibility is that the young woman with the dog is not unlike the reader, an ordinary individual who supports animal research. A final interpretation, as evidenced by the placement of a dog in the picture, may be that supporting animal research represents a very difficult moral decision. Often we are

Figure 4.2 Front cover of Coulston Foundation brochure.

Figure 4.3 Inside page of Coulston Foundation brochure.

forced to choose between what is in the best interest of humans and the well-being of other animals whom we care about. In this way, researchers acknowledge the difficulty and magnitude of the decision when deciding to support animal research.

Frames that deal explicitly with primates move from the general to the more specific. Whereas "animal research" (as evidenced by the example above) is a broad concept involving all species of animals, primate testing deals only with primates. As previously stated, primate testing is a harder sell than testing on mice or rabbits because of primates' evolutionary

Figure 4.4 New Iberia Research Center pamphlet ("Primate Resources").

closeness with humans. Many of the frames employed in favor of primate testing stress progress and humane treatment, thus also drawing upon status quo frames from the medical field. One particular pamphlet begins by showing a young mother with her baby—"This mother and child owe a great deal . . . "—and then moves to the inside of the brochure, where a list of medical accomplishments involving primate research is reviewed.

The brochure concludes with the continuation of the opening headline, indicating to whom the human mother and child "owe a great deal." Above an image of "Donna and son Spudnut," a mother primate and her child, the header reads "to this mother and child."

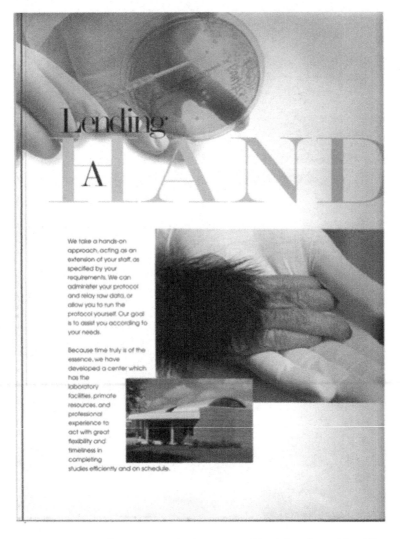

Figure 4.5 New Iberia Research Center pamphlet ("Lending a Hand").

This scientific progress frame is very effective not only in highlighting the scientific arguments but also in invoking positive emotions for the reader. Therefore, the frame that emerges is a scientific progress frame that accentuates discovery and knowledge through the use of primates.

Because treatment also plays an important role in how individuals evaluate primate testing, supporters of primate testing also advance frames that focus on the humane treatment of primates in research laboratories. A pamphlet for the New Iberia Research Center shows that primates are indeed treated well. The images paint a comforting picture for those who may have reservations or be uncomfortable with the use of primates for research.

One image shows a lab assistant bottle-feeding a primate as a mother would her child, and the pamphlet ends with the image of a primate's hand over a researcher's hand as a way to suggest that primates are collaborators with humans in the discovery of treatment and cures.

If there was any concern about whether primates were treated well in laboratories, this "positive treatment frame" should leave little doubt that they are.

MEDIA FRAMES AGAINST PRIMATE TESTING

In response to the organizations that promote and defend primate testing, a growing number of groups are actively working to reduce and eliminate primate testing. Many of the frames used by primate-testing opponents feature a consistent theme of suffering. Almost all the frames that were examined included at least some reference to suffering, with most frames focusing entirely on suffering. For instance, a pamphlet put out by In Defense of Animals shows the immense suffering experienced by primates used for testing.

The title of the pamphlet is "The Plight of Research Chimpanzees," and its cover shows two chimpanzees huddled next to one another. As the reader opens up the pamphlet, she sees a close-up image of a chimpanzee in a cage staring out hopelessly with dull eyes and no way out, a technique that effectively compares the research laboratory to a prison. As the pamphlet continues, it is split into two sections; one is entitled "An Enlightened Future" and shows two primates in their natural habitat playing, while the adjoining section, referred to as "Versus the Dark Ages," shows a chimpanzee locked in a cage.

The images and words are effective in many ways. First, they contrast primates in the research laboratory with those outside the laboratory. Second, the images depict primates in vulnerable positions (in this case, in a cage) as a way to make individuals uncomfortable with the treatment of research primates. Finally, the frame also suggests that primate-testing opponents can both embrace scientific progress and oppose primate testing, as the phrase "an enlightened future" implies, countering scientific

Figure 4.6 In Defense of Animals pamphlet.

opponents who frequently suggest that progress will halt without primate research.

In sharp contrast to the frames that convey the benefits of experiments and emphasize positive treatment, many of the "suffering frames" put forward by groups opposed to primate testing graphically describe and visually depict the actual experiments that are being conducted on primates. One such example comes from the Primate Freedom Project, whose pamphlet shows several gruesome photographs of primates being experimented on by researchers. Identification of the researchers and descriptions of the research are included, along with words like "starved, burned, implanted with probes, force-fed toxic or corrosive chemicals, forced to inhale chemicals . . . " Consistent with the suffering frame that primate research opponents put forth, the description and depiction of experiments enhances the impression of primate suffering.

Opponents are successful in that they directly challenge the scientific progress frame not only by questioning the validity of the science but also by creating a new frame that shows such science gone awry. The image of primates suffering in research laboratories is a significant discursive move away from both the scientific progress and positive treatment frames, which show researchers working collaboratively with primates for the good of building scientific knowledge. The frames that depict primates suffering

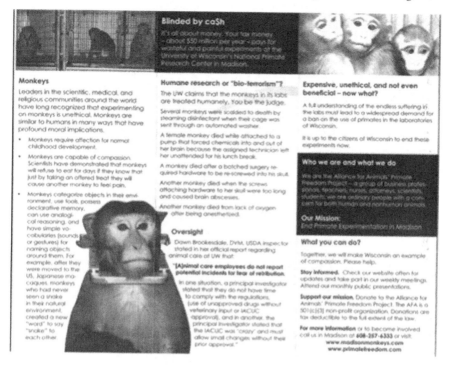

Figure 4.7 Primate Freedom Project pamphlet.

serve two important purposes. First, they assign agency to the individual researchers, in opposition to the status quo frame of presenting the experiments as simply happening to the primates used for research (or implying their participatory consent). Second, and most important for social change, such an "agency frame" helps to account for the ways in which opponents of primate testing frame the capacity for audience members to make a difference, through such means as donations, protests, writing letters to funding agencies and congressional officials and so on.

Besides the suffering frame and agency frame, the other dominant frame that emerges is the "ethics frame." The ethics frame is a direct consequence of the suffering frame. When suffering is shown or implied, and it almost always is, the question becomes whether it is ethical. The frames used by opponents of primate testing makes it quite clear that individuals should place greater emphasis on the ethical dimension. As seen in the Primate Freedom Project's description of what primate testing entails, comparatively less emphasis is placed on the researchers' actual goals and motives than on the actual ethics of the testing. A pamphlet promoting the creation of a primate-testing museum nicely illustrates the ethical frame.

To begin, the pamphlet cover shows a primate behind a cage with the rhetorical question, "How like us need they be?" In doing so, it compels

Figure 4.8 National Primate Research Exhibition Hall pamphlet.

individuals to consider the similarities between primates and humans. Within the pamphlet is a discussion of the similarities between primate testing and the Jewish Holocaust. Pictures of Holocaust victims are juxtaposed with primates undergoing experiments. Linking primate testing with the Holocaust is considered very controversial because the Holocaust is regarded as one of the most violent and ethically shameful moments in human history. Yet, by doing so, the frame does many important things. First, it challenges and debunks the humane treatment frames that are promoted by supporters of primate testing. Like the Holocaust, the treatment of primates cannot be humane and inhumane at the same time. In other words, there can be no in-between. Second, and more indirectly, the Holocaust conjures up images of bad science. Most individuals are familiar with the Nazi scientists who forcibly experimented on Holocaust victims to satisfy scientific curiosities. Showing primates who have no choice in being experimented upon makes a comparison between primate research and these Nazi experiments.

A final technique is using the ethics frame to challenge the scientific progress frame and turn it upside down. This is done by including quotes by well-known scientists describing the advanced cognitive abilities of primates alongside images of primates suffering. This in turn serves to elevate primates to a more humanlike status. In this way, scientists' own words are used against them to highlight the ethics of experimenting on primates. Clearly, then, the ethics frame plays a very important role in many of the messages put forth by those working to stop primate testing.

CONCLUSION

I began this chapter with the goal of better understanding how supporters and opponents frame the primate-testing controversy. To this extent, I investigated a small but representative subset of pamphlets put out by the different groups working for and against primate testing. There are some interesting takeaways from this analysis that can further our understanding of how the primate-testing controversy is framed and what it means for understanding animal-human communication.

At first glance, it would seem that supporters and opponents of primate testing have little in common in how they think about primate testing. To be sure, the two sides do not share goals, and Deborah Blum, among others, has already shown quite candidly that they do not communicate well with one another. Yet, regarding the frames advanced by both sides, it would seem unfair to draw the conclusion that both sides are dramatically different in how they frame the issue. In fact, a more cautious conclusion seems to be that both sides appeal to a similar set of core principles and are much closer to one another than what might have originally been thought.

A theme that appears in both sets of frames is the idea of suffering. In the case of message frames that support primate testing, suffering is more implicit and must be drawn out by the audience member. A scientific progress frame makes scientific achievement and discovery the central focus of the message but it also suggests that the absence of scientific progress could delay or impede future progress, which in turn would lead to greater suffering for humans. This is best illustrated in the pamphlet that juxta-poses a primate mother and child with a human mother and child. It should be inferred by the audience member that in order for the human mother and child not to suffer, the primate mother and child must experience suffering, or at least undergo experimentation. In this way, the frame acknowledges the reality of suffering but gives audience members the choice of whom they would prefer to see suffer, humans or primates. Of course, the use of a positive treatment frame by proponents of primate testing downplays the reality of suffering. For supporters, the choice is obvious and reinforces the tangible divisions between humans and primates. Supporters have a relationship with primates of deep appreciation and gratitude, extended from the human to the nonhuman primate. By contrast, in message frames from the groups opposing primate testing, suffering is explicit and often the main focus of the message content. Suffering is observed through what the primates endure, the laboratory conditions and the graphic detail in which the messages appear. Audience members cannot escape the presence of suffering and are confronted with images that are intended to make them question whether it is reasonable to inflict great amounts of suffering upon beings who are, despite their differences from humans, similar to them in many respects. Both sides address suffering but focus their attention on dif-ferent aspects of suffering.

Another area of overlap concerns the concept of progress. Frames pro-moting primate testing emphasize the progress that has been made through primate testing and the potential for continued and even greater scientific progress. Some frames explicitly mention some of the scientific and medical breakthroughs that have occurred as a result of primate testing, while other frames criticize the opponents for blocking progress. By contrast, frames opposing primate testing focus on the lack of progress in two important ways. First, they question the validity of the science and its ability to pro-duce accurate results. Often, these frames are accompanied by a quote from someone in the scientific research community who is critical of pri-mate testing. These frames reappropriate the idea of scientific progress by using ethical standards and relevant scientific knowledge regarding primate social and cognitive behavior to promote "an enlightened future." They effectively characterize primate researchers' work as trivial and detached from common and ethical sense. A lack of progress is also conceptualized as the moral indifference of people and their lack of thought or care for other primates in favor of human selfishness. Many of these frames use terms like "torture" or "dark ages" to suggest that individuals' social atti-tudes need to catch up with the times.

In their frames, the two sides do differ dramatically in their selec-tion of images involving primates. Supporters of primate testing tend to depict primates in comfortable and content states, usually socializing with other primates. Primates are depicted in open spaces with plenty of social enrichment to keep them occupied. Moreover, primates are seldom shown in cages, no doubt a conscious effort to keep individuals from equating a laboratory with a prison. When "experiments" are shown in pictures, they usually show a researcher innocuously measuring a heartbeat or taking a pulse with a primate who looks eager and willing to cooperate. These types of images complement the positive treatment frame, in which researchers are seen as compassionate in their care for and work with primates.

Unlike the supporters' brochures, which feature idyllic settings and lack images of primate testing, the brochures opposing primate testing frequently rely on a set of pictures that paint a very different and disturbing portrait of what is happening inside the laboratory, or "behind bars." Many of the pictures show images of primates experiencing invasive procedures, such as being strapped in chairs, being forced to breathe toxins and having their heads cut open by researchers. The pictures almost always show primates in helpless states beyond their control or experiencing uncontrollable fear and hysteria.

So gruesome and unpleasant are the photos involving primates inside laboratories that researchers have taken several steps to suppress such images, including restricting their publication in scientific journals, increasing security around national primate center facilities and even destroying videotapes when they are requested by activists through the Freedom of Information Act (Leuders). In contrast to the animal research industry's

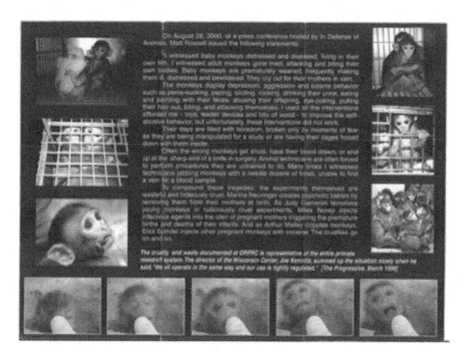

Figure 4.9 National Primate Research Exhibition Hall pamphlet (inside).

images of primates in pleasant settings acting as collaborators in the scientific research process, these images are likely to evoke anger and disgust, as well as compassion for primates. Such emotions make the reader question not only whether treatment meets a high threshold of care but also whether use of primates is ever appropriate. Clearly, the strategic use of images plays a significant role in stirring emotions and in many ways simplifies the primate-testing controversy by shifting it from a discourse-oriented debate (i.e., primate testing is beneficial because it results in scientific progress) to a more visually oriented debate (i.e., gruesome images) in which audience members are repulsed by what they see and consequently moved to action.

An important contribution of this study is that it describes a particular communicative phenomenon that has yet to be fully explored. In this instance, how primate testing is framed in pamphlets is described with an emphasis on the different words, terms, phrases and images that are used in the debate. As we consider frames in the primate-testing controversy, we see two groups with obviously different goals but with striking similarities in how they frame the debate. One group is trying to maintain the status quo and reinforce the notion that scientific progress is beneficial and that primate testing is worth the sacrifice, whereas the other group attempts to challenge the status quo by reframing what might be the dominant discourse (e.g., science is cold and uncaring, progress comes at a cost, etc.) in order to advance their interests. In each case, both groups appeal to a certain set of core values that they believe will resonate with audience members. Thus, how the issue is framed within the core value should matter in how individuals think, and ultimately make evaluations, about primate testing.

While description is a useful first step in identifying the type of content that one is observing, very little if anything can be said about what kind of impact such content might have on individual-level thoughts, attitudes and behaviors. Future studies in communication scholarship and other social science disciplines should focus on different types of effect variables to see what influence such messages might have on audience members.

Many of the message frames that were investigated in this study came from advocacy groups and biomedical research companies who were sending material to their supporters or the 'already converted' who require little, if any, persuading in adopting their views and positions. As of now, we still know little about the individuals in the public sphere who support and oppose primate testing. Future studies could construct a profile of those individuals by identifying the key variables that distinguish them from one another, including the variables that go beyond simple demographic-level information. Additionally, to get a better sense of how effective these messages are, some possibilities might be to manipulate some aspect of the message design (message strength, image or source) to see how the level of support for primate testing is impacted. Survey research would also be beneficial because it would be able to shed greater light on how individuals feel about primate testing. At the same time, it would be able to capture a large, diverse and representative sample, often absent in experimental research.

A final point of emphasis for future research would be for communication scholars to better understand the mental processes related to the primate-testing debate and how they extend to other areas of human-animal communication. For example, how do those who oppose primate testing think about other issues involving animals, like intensive agriculture, animals in entertainment, the exotic pet trade and so on? Such studies would not only further our understanding of the psychological processes underlying human communication and the power of framing to shift consciousness but would also help us make better sense of the complex relationships people have with animals. Public outrage over Michael Vick's dog-fighting scandal, mounting concern over factory farm practices, increased demand for "humanely raised meat," opposition to commercial whaling, a shift away from animal fabrics such as fur and leather and opposition to animal testing all suggest that debates that orient us toward the complexities of human-animal interaction offer a promising and timely area of communication research.

Finally, the issue of why primates are important deserves mention. One thing that seems abundantly clear is that both groups value primates but have dramatically different reasons for doing so. Despite the moral dilemma that many primate researchers insist they experience when performing vivisections on primates, researchers value primates solely for their physiological and psychological similarities to humans, with almost complete devaluation of and disregard for their capacity for rich mental and social lives. Misery in laboratories is commonplace among primates, who exhibit stereotypical behavior that includes endless pacing, spinning and self-mutilation for years on end (Reinhardt and Rossell). Those who defend primate testing are hard-pressed to come up with strong and cogent ethical reasons as to why it would not be ethically problematic to subject human primates (including children) to experimental tests that routinely involve the infliction of excruciating pain through various means of invasive procedures such as incisions, injections, infections, illnesses and a host of other injuries. Replacing a human subject in such an experiment with a nonhuman primate, however, becomes not only ethically justifiable and desirable but is publicly defended by researchers and laypersons alike, publicly lauded and financed and the recipient of continuing federal and state support.

Prior justifications showing human intellectual and emotional superiority over animals have become more difficult to maintain in light of increasing evidence showing the unique and intricate cognitive and emotional lives of animals, especially those of primates whose lives have been the subject of fascination for centuries (Balcombe). Few people can deny any longer that there is immense pain and suffering involved in primate testing; thus, the question becomes whether society as a whole is ready to forego some possible scientific and medical benefits from primate testing and recognize the inherent self-worth of primates as beings deserving of their own lives and experiences. Given the historic trajectory of humans extending basic courtesies and rights to other groups (Bogle 53), it seems inevitable that primate testing should come to a permanent end some day. This process, though, can be accelerated by artfully shaping the frames used in the debate. Specifically, the ethics frame represents the best possibility for ending primate testing since it speaks to a higher set of values than even the scientific progress frame. The very recent example of the United States apologizing to the Guatemalan government for syphilis experiments unknowingly conducted on prisoners, mental patients and soldiers in the late 1940s is a stark reminder that ethics trump science ("U.S. Apologizes"). In other words, we as a society place more emphasis on ethics than we do on science. Therefore, opponents of primate testing should continue to stress the ethics frame and shift the debate from the possibility of scientific progress to the ethics of using primates for experimental research. If humans continue to take seriously their relationships with animals, the debate over primate testing and the fates of thousands of captive primates may hinge on which side is best able to frame the issue.

REFERENCES

Bailey, Jarrod, Andrew Knight, and Jonathan Balcombe. "Chimpanzee Research: An Examination of Its Contribution to Biomedical Knowledge and Efficiency in Combating Human Diseases." *Project RandR: Release and Restitution for Chimpanzees in U.S. Laboratories: A Campaign of the New England Anti-Vivisection Society* (2007): 1–47. Print.

Balcombe, Jonathan. *Second Nature: The Inner Lives of Animals*. New York: Palgrave Macmillan, 2011. Print.

"Biomedical Research Group Launches Groundbreaking TV Campaign" Reuters. 16 Sept. 2009. Web. 16 Jun. 2011.

Blum, Deborah. *The Monkey Wars*. New York: Oxford University Press, 1995. Print.

Bogle, Rick. "Animal Experimental and Human Rights." *Human Rights Review* 4.2 (2003): 53–61. Print.

Cantlon, Jessica F., and Elizabeth M. Brannon. "Basic Math in Monkeys and College Students." *PLoS Biol* 5.12: e328. doi:10.1371/journal.pbio.00503282007. 18 Dec. 2007. Web. 19 Feb. 2012.

Contopoulas-Ioannidis, Despina G., Evangelia Ntzani, and John P. A. Ioannidis. "Translation of Highly Promising Basic Science Research into Clinical Applications. *American Journal of Medicine* 114.6 (2003): 477–484. Print.

D'Angelo, Paul. "News Framing as a Multiparadigmatic Research Program: A Response to Entman." *Journal of Communication* 52.4 (2002): 870–888. Print.

Druckman, James N. "The Implication of Framing Effects for Citizen Competence. *Political Behavior* 23.3 (2001): 225–253. Print.

Egan, Louisa, C., Paul Bloom, and Laurie R. Santos. "Choice-Induced Preferences in the Absence of Choice: Evidence from a Blind Two Choice Paradigm with Young Children and Capuchin Monkeys." *Journal of Experimental Social Psychology* 46.1 (2010): 204–207. Print.

Entman, Robert M. "Framing U.S. Coverage of International News: Contrasts in Narratives of the KAL and Iran Air Incidents." *Journal of Communication* 41.4 (1991): 6–27. Print.

Entman, Robert M. "Framing: Toward a Clarification of a Fractured Paradigm" *Journal of Communication* 43.1 (1993): 51–58. Print.

Francione, Gary L. "The Use of Nonhuman Animals in Biomedical Research: Necessity and Justification." *Journal of Law, Medicine, and Ethics* 35.2 (2007): 241–248. Print.

Institute of Medical Progress. "Chimpanzees in Biomedical and Behavioral Research: Assessing the Necessity." 15 Dec. 2011 Web. 4 Jan. 2012.

"Gallup Poll Puts Majority Behind Research." *Speaking of Research.* 4 Jun. 2010. Web. 22 Jun. 2011.

Gamson, William, A. and Andre Modigliani. "The Changing Culture of Affirmative Action." *Research in Political Sociology*, Vol. 3, Greewich, CT.: JAI Press, 137–177. Reprinted in *The Political Sociology of the State*, ed. Richard G. Braungart and Margaret M. Braungart, Greenwich, CT: JAI Press, 1991, and in *Equal Employment Opportunity*, ed. Paul Burstein, New York: Aldine de Gruyter, 1994.

Goldberg, Alan M. "Animals and Alternatives: Societal Expectations and Scientific Need." *ATLA: Alternatives to Laboratory Animals* 32.6 (2004): 545–551. Print.

Greek, Ray C., and Jean S. Greek. *Sacred Cows and Golden Geese: The Human Cost of Experiments on Animals.* New York: Continuum, 2002. Print.

Jenness, Valerie. "Social Movement Growth, Domain Expansion, and Framing a Social Problem." *Social Problems* 42.1 (1995): 145–170. Print.

Kinder, Donald R., and Lynn M. Sanders. "Mimicking Political Debate with Survey Questions: The Case of White Opinion on Affirmative Action." *Social Cognition* 8.1 (1990): 73–103. Print.

LaFollette, Hugh, and Niall Shanks. *Brute Science: Dilemmas of Animal Experimentation.* New York: Routledge, 1997. Print.

Lakoff, George. *Don't Think of an Elephant!: Know Your Values and Frame the Debate.* White River Junction, VT: Chelsea Green, 2004. Print.

Leuders, Bill. "Isthmus Daily Page." Isthmus, 11 Aug 2006. Web. 22 Jul 2010. <http://www.thedailypage.com/isthmus/article.php?article=2012>.

McLeod, Douglas M., and Benjamin H. Detenber. "Framing Effects of Television News Coverage of Social Protest." *Journal of Communication* 49.3 (1999): 3–23. Print.

"More than a Makeup Trend: New Survey Shows 72 Percent of Americans Oppose Testing Cosmetic Products on Animals." *PCRM.* Autumn 2011. Web. 18 Jan. 2012.

National Institute of Health. "NIH Research Involving Chimpanzees." 21 Dec. 2011. Web. 4 Jan. 2012.

Pan, Zhongdang, and Gerald M. Kosicki. "Framing Analysis: An Approach to News Discourse." *Political Communication* 10.1 (1993): 55–75. Print.

———. "Framing as a Strategic Action in Public Deliberation." *Framing Public Life: Perspectives on Media and Our Understanding of the Social World.* Ed. Stephen D. Reese, Oscar H. Gandy Jr., and August E. Grant. Mahwah, NJ: LEA, 2001. 227–230. Print.

Price, Vincent, David Tewksbury, and Elizabeth Powers. "Switching Trains of Thought: The Impact of News Frames on Readers' Cognitive Responses." *Communication Research* 24.5 (1997): 481–506. Print.

"Public Praises Science; Scientists Fault Public; Media: Scientific Achievements Less Prominent than a Decade Ago." Pew Research Center for the People and the Press. 9 Jul. 2009. Web. 19 Jun. 2011.

Rajala, Abigail Z., Katherine R. Reininger, Kimberly M. Lancaster, and Luis C. Populin. "Rhesus Monkeys (*Macaca mulatta*) Do Recognize Themselves in the Mirror: Implications for the Evolution of Self-Recognition." *PLoS ONE* 5.9: e12865. doi:10.1371/journal.pone.0012865 [pdf]. 2010. Web. 17 Feb. 2012.

Redford, Joshua S. "Evidence of Metacognitive Control by Humans and Monkeys in a Perceptual Categorization Task." *Journal of Experimental Psychology: Learning, Memory, and Cognition* 36.1 (2010) 248–254. Print.

Reinhardt, Viktor, and Matt Rossell. "Self-Biting in caged Macaques: Cause, Effect, and Treatment." *Journal of Applied Animal Welfare Science* 4.4 (2001) 285–294. Print.

Scheufele, Dietram A. "Framing as a Theory of Media Effects." *Journal of Communication* 49.1 (1999): 103–122. Print.

Strum, Shirley C. *Almost Human: A Journey into the World of Baboons*. Chicago: U of Chicago P, 2001. Print.

"U.S. Apologizes for Syphilis Tests in Guatemala." *New York Times* 1 Oct. 2010. Web. 22 Jun. 2011.

Valkenburg, Patti M., Holli A. Semetko, and Claes H. de Vreese. "The Effects of News Frames on Readers' Thoughts and Recall." *Communication Research* 26.5 (1999) 550–569. Print.

VandeBerg, John L., and Stuart M. Zola (2005). "A Unique Biomedical Resource at Risk." *Nature* 437.1 (2005): 30–32. Print.

Williams, Nigel. "More People Accept Animal Research." *Current Biology* 16.1 (2006): R2. Print.

5   Absorbent and Yellow and Porous Is He
Animated Animal Bodies in *SpongeBob Squarepants*

Shana Heinricy

In animation, nonhuman animals have long been the subject of surreal and amazing torture, from Tom and Jerry slicing and dicing each other to the antics of Wile E. Coyote and Roadrunner repeatedly blowing each other up. *The Simpsons* (1989–present) parodies this violence through the characters Itchy and Scratchy, portraying bloody and graphic violence to howls of amusement from Lisa and Bart. The animated nonhuman animal bodies are stretched, chopped, stomped and dissected for the visual pleasure and amusement of the viewer. Of course, this is one of the great pleasures of animation. The impossible is possible and consequence can be nullified. The animated world is usually not completely abstract, but instead takes visual cues from known objects, such as invoking "puppyness" in an animated puppy, even though the representation does not look like a nonmediated puppy. This level of abstraction, indicating an unembodied puppy, thus allows for violence to be committed on its mediated cuteness. The violence is not simply violence, because it is seemingly without consequences. It also evokes wonder at the fantastical possibilities and creates a break from everyday life. In other words, the viewer's pleasure in animated and animal violence is not necessarily sadistic, but instead fantastic. Animation allows for this pleasure in ways that other media cannot, due to the abstraction of representation and the pliability and lack of fixity of animated bodies, especially animated animal bodies.

Nonetheless, this sort of animated violence committed on nonhuman animal bodies, typical of traditional U.S. animation, is characteristic of a culture that condones such violent acts on sentient, unmediated animals. Representation is a means of cultural control over the conception of animals. Animated representations help to mirror and constitute animal bodies in American culture, particularly reinforcing the dichotomy that has been created between companion animals and worker animals, which renders the suffering of worker animals invisible. Worker animals produce a service or product for human culture, such as rescue dogs, milking cows, pigs raised for meat, horses that are ridden, animals in zoos and chickens' eggs and bodies. Companion animals exist for the pleasure of their humans. Somehow, people are able to separate their love of pets from the systematic torture and slaughter of animals for food and products. It is acceptable in society in the United States to call oneself "an animal lover" while funding, participating in and endorsing systems of torture for "food" animals.

Visual images communicate ideologies to viewers. They may be difficult to pinpoint, difficult to express in words because of their visual form, but they are nonetheless present as a form of communication. It is not the inconsistency between slaughtering animals for food and loving pets that I question, but the fact that food production in the United States no longer simply requires the slaughter of animals that have lived in otherwise healthy conditions. Instead, the interests of animals and the suffering of their bodies are ignored in order to increase profits of factory farms. It is beyond the focus of this paper to explain the extent of the mistreatment of farm animals or to convince the reader of the abuse. I want to point out, however, that the dichotomy between pets and other animals results in the exploitation of many animals for capitalist goals and in material harms to animal bodies. Jonathan Burt offers an explanation of how humans resolve such dissonance:

> One of the most commonly noted examples of this is the inconsistency between the celebration of animals in, say, nature films or pet-keeping and the simultaneous sanctioning of, or presiding over, the destruction of animal life on an enormous scale. It is, perhaps, to avoid the full implications of this strangeness that we choose to find a place for them in a well-organized cultural logic that divides the animal world into categories like pet, vermin, threatened species, and expendable species. (Burt 204)

This dichotomy has been created, maintained and rendered invisible, in part, through the representations of cartoon animal bodies in the United States, which the medium of animation helps to make possible. First, I argue that representation is a means of power and control. I explain the ways that animal representations have been used to connote Otherness and disparage humans in order to show a connection between racism, sexism and speciesism. Next, I illustrate how animals have been linked to animation since the origins of American animation, making it an important place to study the representations of animals. Finally, I provide a case study of the popular animated television series *SpongeBob Squarepants* (1999–present) as a specific example of the ways in which the visual representation of animated animal bodies helps to create, maintain and render invisible speciesist ideologies.

ANIMAL REPRESENTATIONS

The interests of animals have been marginalized in Western societies. The animal rights debate is a heated one in the animal defense community, due in part to the patriarchal construction of "rights" discourses that are based on the founding fathers' conception of rights (Adams 13). However, the capacity of animals to suffer indicates that they have interests, if not rights, namely their interest in not suffering (Singer, "Animal Liberation or Animal Rights?" 16). The interests of animals have been ignored. Without detailing the horrors of factory farming and testing procedures on animals, it is fair to say that animals are frequently treated in cruel and inconsiderate ways, from dog-fighting rings to chicken coops so crowded that chickens are debeaked so they do not peck each other to death. People for the Ethical Treatment of Animals (PETA) details the atrocities experienced by animals used for food, clothing and experimentation at peta.org/issues. This lack of consideration of animals' interests is evidence of "speciesism." According to Peter Singer, speciesism is "a prejudice or attitude of bias in favor of the interests of members of one's own species and against those of members of other species" (*Animal Liberation* 6). In this way, he considers speciesism a form of prejudice akin to racism and sexism, intertwined in one system of dominance.

Although instances of speciesism may be traced throughout history, many scholars argue that society is currently experiencing an increased subjugation of animals that is a result of modernity, closely linked to the progression of capitalism. Discourses of God and nature have been displaced in the modern era to place humans and human interests in production at the center of the discourse (DeLuca 54). John Berger states in his essay "Why Look at Animals?" that the separation between humans and nonhuman animals is symptomatic of the urbanization of the Western world. As people moved to the city, they were no longer involved with animals. The use of animals for labor and for food was rendered invisible to those in the cities. As industrialization accelerated, even those in relatively rural environments increasingly purchased, rather than produced, food. In addition, urbanization decreased contact with wild animals, which made animal corpses (food) and pets the common ways for people to interact with animals. He argues that pets are not "real" animals because a true animal exists separate from humans (256). While this notion that pets are not animals is problematic and shows a romanticizing of rural environments, it is important to point out that humans' constructions of pets often have little to do with the animals, and more to do with the interests and personality of the owners, who attribute anthropomorphized characteristics, motivations and emotions to their pets. In other words, the rhetorical construction of "pets" may have little connection to the lived lives of animals and instead says more about how people would like to view animals. Gradually, the sign of 'animal' lost its traditional signifier in modernity and has been replaced with a new meaning that is disconnected from animal bodies.

Animation is a clear example of the lack of connection between the sign and signifier of 'animal.' Part of the fun, or play, of animation is the ability of the medium to attach different signs and signifiers. For example, the image of chair may not simply be a chair, but instead an insane machine intent on eating the main character's sandwich. The normal fixed relationship between image and meaning is broken down in animation, creating humor and wonderment. A substantial way in which this is done is through the use of animals, attributing unexpected signifiers to the sign of 'animal.' For example, an anthropomorphized skunk named Pepe Le Pew is depicted as a diehard romantic. Of course, he still exhibits a sort of comic stupidity, not realizing the feline object of his affections is not a skunk.

Representations of animals are dissociated from the interests of animals, as skunks are most likely not interested in the human conception of romance. People refer to their food in terms that dissociate it from the animal body being eaten (e.g., pork or beef instead of pig or cow). In addition, animals' individuality is eliminated through language to further distance any discussion of animals' interests. For example, "meat" is what Carol Adams refers to as a mass term. Meat has no quantity. It erases the fact that "meat" is made up of many individual animals, each with likes and dislikes (*Neither Man Nor Beast* 27–30). The *Silly Symphonies* (1928–1939) cartoon "The Three Little Pigs" shows an image of "dad" as a butcher's chart, with the various parts of the body marked up and named for butchering, which draws a comic comparison between the animated world of the pigs and the nonmediated one.

I do not want to argue that animals have some sort of "actual" essential animal nature that representations miss. Animals differ from representations of animals in that they can exist apart from humans. Animal representations are solely the constructions of humans, thus allowing humans to control conceptions of animals. The animals about which I write may or may not have any relation to humans. They may have interests, attitudes and likes and dislikes that we may not ascribe to that particular animal via representations (such as a cat who likes to swim). Animals can exist apart from humans, eating what and when they choose, getting pleasure from whatever activities they choose. Adult animals are, after all, adults. Whereas pets and wildlife representations tend to portray animals as infants, thus needing the care and protection of people, the reality is that most adult animals would survive and raise their own young competently without human interference. U.S. animation consistently uses neoteny in order to make animals appear cute, likeable and subservient. Such representations of animals allow humans to exert control over the image of animals in ways that benefit human interests.

W. J. T. Mitchell notes that "human power over others is secured by mastery of representations" (332). Through the representation of animals in Western culture, humans have sought to return a signifier to the sign of 'animal.' However, this signifier has been constructed to erase the systematic torture and slaughter of animals in factory farms and testing environments, therefore allowing humans to continue their control of, and profit from, animals. In essence, lab animals and animals in factory farms have been erased from the consciousness of Western society. Even films such as *Babe* (1995), which on some levels critique the practice of using animals for food, romanticize farm life. The farm on which Babe lives is clean and allows all of the animals to roam freely, eating as they please; it erases the reality of the factory farms where most farm animals live.

The representation of animals does not tell viewers simply about animals, but also about humans. Verbally, animal terminology has been used to disparage humans. The term "beast," while associated with the animal world, provides more insight into what humans dislike about other humans than it does into the nature of animals. As Carol Adams states, "The concept of the beast exists to be self-referential, as a comment on humankind. Animals actually are neither man nor beast—neither mere caricatures of their own lives, nor stupid—but, like human beings, animals with social needs and interests" (*Neither Man Nor Beast* 13).[1]

Clearly, the category of animal has been used for social othering. According to W. J. T. Mitchell, "Animals stand for all forms of social otherness: race, class and gender are frequently figured in images of subhuman brutishness, bestial appetite, and mechanical servility" (333). In the Americas, African slaves were termed beasts because they were deemed closer to animals than white men due to their lack of "progress" and "industrialization." In a similar way, animal societies were unfairly judged according to human standards for societies. At the same time, women were deemed beasts due to their lack of intelligence, and have been termed "pieces of meat" and "chicks," which illustrates the link between consuming animal bodies and the sexual consumption of women (Adams, *The Sexual Politics of Meat* 42).[2] In this chapter, I explore how animated representations of animals enable humans to continue disregarding the interests of animals. Animal images have little to do with nonmediated animals, and more to do with human needs and desires.

ANIMATION AND THE ANIMAL

Animals and animation have been so intrinsically tied together since the birth of Western animation that I must wonder if the animal images in animation led, in part, to the disregard of the medium of animation. As Steve Baker stated, "The animal is the sign of all that is taken not-very-seriously in contemporary culture: the sign of that which doesn't really matter" (174). In the United States, both animation and animals have long been associated with children. Animation has had a difficult time establishing itself as an art form, and is often disparaged as an art form (Stabile and Harrison 2). It has even been deemed insulting to render certain events, such as the Holocaust, in animation or comics, because animation and comics are viewed as flippant media (Baker 140). The medium of animation itself, regardless of content, is somehow perceived to be frivolous and unintelligent, not for 'serious' matters. It is telling, therefore, that animals are considered an appropriate subject matter for animation. Since the appearance of Felix the Cat, Mickey Mouse, Bugs Bunny and many other animal characters in the golden era of American animation (Bendazzi 55, 63, 97; Wasko 9), animals have played a central role in people's understanding of animation in the United States.[3]

Likewise, animals were part of animation since its origins. McKay's *Gertie the Dinosaur* (1914) and *The Story of the Mosquito* (1912) are two examples of the original works featuring animated animals, with the mosquito undergoing perhaps the earliest instance of animated animal violence when it explodes at the end of its story from gorging on too much blood. This plot is repeated with John Randolph Bray's *Artist's Dreams* (aka *The Dachshund and the Sausage*, 1913), when the Dachshund similarly explodes due to sausage consumption. In the late 1920s, John Randolph Bray put out the first color animated film, *The Debut of Thomas Cat*, which featured an animal lead character. *Felix the Cat* (1920–1928), created by Otto Mesmer, likely featured the first animated animal to assume celebrity status, followed soon after by Mickey Mouse with the success of Disney's *Steamboat Willie* (1928), the first widely distributed synchronized sound cartoon. With Mickey Mouse came a plethora of his animal friends, including Minnie Mouse, Goofy, Donald Duck and Pluto. Disney's *Silly Symphonies* series (1928–1939), animated shorts based on a musical composition, featured many animal themes, with the animals often appearing goofy as they move to music, solidifying the "silly animal" motif in U.S. animation. Walter Lanz's *Andy Panda* (1939–1949) introduced the infamous and long-lasting *Woody Woodpecker* (1940–1972). Ub Iwerks added the eponymous character of *Flip the Frog* (1930–1933) and Charles Minz added those of *Oswald the Lucky Rabbit* (1927–1928) and *Krazy Kat* (1916–1929) to this ever-increasing menagerie of animated animal stars. Paul Terry created *Gandy Goose* (1938–1955) and *Kiko the Kangaroo* (1936–1937) before creating an animal star of his own with *Mighty Mouse* (1942–1961).

As animals became established characters in early U.S. animation, Warner Brothers made comic geniuses out of animals, in the form of *Looney Tunes* and *Merry Melodies* characters such as Bugs Bunny, Pepe Le Pew, Speedy Gonzalez, Porky Pig, Daffy Duck, Roadrunner, Wile E. Coyote, Tweety, Sylvester and Foghorn Leghorn, with cartoons which ran from the mid-1930s until the late 60s. Part of the humor was achieved through harming the animals' bodies in exaggerated ways, such as dropping an anvil on one of their heads. The character becomes dazed and confused by the event, perhaps crumpling like an accordion under the weight of the anvil, but ultimately remains unharmed, showing that violence has no consequences on the animal's body. MGM's *Tom and Jerry* (1940–1967), created by William Hanna and Joseph Barbera, further perfected the humorous and violent animal motif. Most of these Warner Brothers and Hanna-Barbera cartoons were reaired as television series for a new generation.

While much of television animation was recycled from earlier cartoons, full-length animated television series became more prevalent in the 1960s, many of which featured animals and are documented in Jeff Lenburg's encyclopedia. Shows such as Jay Ward's *Rocky and His Friends* (1959–1961) and *The Adventures of Hoppity Hooper* (1964–1967), the Academy Award–winning *The Pink Panther Show* (1969–1970) and Hanna-Barbera's *Scooby Do, Where Are You?* (1969–1972) were reincarnated as spinoffs from the original series. The 1980s saw a plethora of animal-based animated television series, many of them toy-based in order to exploit marketing opportunities. Examples include *Alvin and the Chipmunks* (1983–1990), *Muppet Babies* (1984–1990), *My Little Pony 'n Friends* (1985–1986), *Thundercats* (1985–1987), *The Berenstein Bears* (1985–1987), *The Care Bears* (1985–1988), *Chip 'n Dale Rescue Rangers* (1988–1989), *Count Dukula* (1988–1993), *The Teenage Mutant Ninja Turtles* (1987, 1990–1996) and *Darkwing Duck* (1991–1993).

Animals were connected with racial stereotypes in early animation; for example, an ape was used to represent a black character. In *Saturday Morning Censors*, Heather Hendershot argues, "Historically, cartoon animalization has been a strategy to mask overt (and therefore potentially controversial or censorable) ethnic stereotyping" (102). Rather than attempting to add racial characteristics to the features of Speedy Gonzalez and Slowpoke Rodriguez, the animators added stereotypical Mexican pauper attire. Among the the half-human, half-cat characters of *Thundercats*, Panthro (literally a black panther) was colored darkly and given stereotypical African-American features, such as a wider nose and larger lips than the other characters. Animals, rather than humans, allowed for more deniability regarding racial stereotypes. Panthro, a Thundercat coded as black, was not black because it would be ridiculous to call a cat African-American. The industry claimed that animals did not have ethnicities. Therefore, race and animality are inextricably tied in animation.

In the 1990s, with the return of primetime animation with *The Simpsons*, animation turned toward cartoons focusing on human main characters. Human main characters always existed in television animation, such as in Hanna-Barbera's *The Flintstones* (1960–1970), *Popeye* (1960) and *He-Man and Masters of the Universe* (1983). However, they were surrounded by large numbers of animal-based cartoons, such as those listed above. In the 1990s, adult-oriented cartoons became prevalent, all of which featured humans as their characters, such as *The Critic* (1994–1995), *The Simpsons* (1989–present) and *South Park* (1997–present). Again, it is interesting that animal main characters were removed to create animation that could be taken seriously by many adults. Disney films, in the era of "New Disney" under Eisenstein, provided animal sidekicks as comic relief to human main characters, rather than an all-animal cast, with the exception of *The Lion King* (1994). In addition, most of the popular animated television series of this period focused on humans, such as MTV's *Clone*

High (2002–2003), Craig McCracken's *The Powerpuff Girls* (1998–2004), Genndy Tartofsky's *Dexter's Lab* (1996–2003), Arlene Klasky's *Rugrats* (1991–2004) and *Rocket Power* (1999–2004) and Craig Bartlett's *Hey Arnold* (1996–2004)—this list could go on and on.

The 1990s also saw a shift from animated animal casts to "socially progressive" animation with human casts. This was especially the goal of networks aimed at children, such as Nickelodeon and Disney Channel. This progressiveness manifested itself, as evidenced by the contemporary shows themselves, by removing much of the violence that was associated with animation and including female main characters who appear multifaceted and independent, including representations of healthy nontraditional families and including more nonwhite characters. The skin of these nonwhite characters is shaded darker than that of other characters; however, their race is never mentioned within the shows, human voices are rarely ethnically coded and racial issues are not acknowledged in any way. *SpongeBob Squarepants* (1999–present), as one of the few series on animated television featuring an all-animal cast, does not give any indication of characters' race, although the voice actors are all white. By the early 2000s, *SpongeBob Squarepants* was the top-rated animated program, with an entirely underwater animal cast.

SPONGEBOB SQUAREPANTS

The creation of Steve Hillenburg, *SpongeBob Squarepants* first aired on Nickelodeon in 1999 and is still on the air, completing its ninth season in 2012, according to the online Internet Movie Database (IMDB). Hillenburg used his bachelor's degree in marine biology and master's degree in experimental animation from California Institute of Arts to create an animated series modeled after a Polynesian tidal pool. The top-rated children's show in 2002, *SpongeBob* averaged 56 million viewers per month in that year (Rosenthal 45). A third of the viewers were over 18 years old, indicating the program's popularity with multiple age groups (Johnson D1; Rosenthal 45). In addition, *SpongeBob* merchandise brought in more than $500 million in 2002 (Rosenthal 45). SpongeBob graced the big screen for the first time in 2004 with a film that grossed $85.4 million domestically and $54.7 million abroad. A second film is slated to come out in 2014 (Chozick B4). Despite being surrounded by animation focusing on human characters, *SpongeBob Squarepants*'s success illustrates that animal animation is still both relevant and profitable. Moreover, precisely because *SpongeBob* is so popular, it has the power to shape human perceptions and understandings of marine animal life.

Nickelodeon is a "kids" network, meaning that it specifically targets children. It claims to be concerned with adults only to the extent that adults will allow their children to watch Nickelodeon (Keveney 1E). This is different from Cartoon Network, which seeks to provide animation of interest to both children and adults, targeting adults specifically with late night blocks such as "Adult Swim" (Major 1; Werts D10). Ironically, while focusing on children, Nickelodeon produced two programs that were extremely popular with adults, both reaching a cult classic status: *Sponge-Bob Squarepants* and the now defunct *Ren & Stimpy*. Part of the adult appeal of *SpongeBob* may be its use of celebrity voice actors whose voices are known to adults (Marion Ross from *Happy Days*, Bill Gagerbakke from *Coach*, John O'Hurley from *Seinfeld*, Ernest Borgnine and Tim Conway; Johnson D1).

SpongeBob focuses on life in Bikini Bottom, which is an underwater community of sea creatures. The main character, SpongeBob Squarepants, is a hardworking fry cook at the Krusty Krab restaurant. As his name denotes, he is a sponge, but not a natural-looking ocean sponge. No, he is a rectangular, yellow kitchen sponge. His character is "square" both physically and metaphorically. He is kind, considerate, naïve and somewhat geeky, not quite as cool as the body-building fish on the beach. For example, in the episode "No Weenies Allowed," a bouncer at a "tough" bar sent SpongeBob to "Super Weenie Hut Jr.'s" in an effort to show that he was simply not suitable for the "tough" bar. In addition to his geeky naïvety and kindness, SpongeBob is devoted to his work, and quite skilled at making Krabby Patties and blowing bubbles. His boss, Mr. Krabs, is a large, tight-fisted red crab as well as an ex-sailor. SpongeBob's efficiency results in large amounts of money for the restaurant and Mr. Krabs.

As the theme song states, "he lives in a pineapple under the sea." Squidward and Patrick are two major characters who live next to SpongeBob's pineapple. Patrick is a pink starfish who wears swim trunks and is Sponge-Bob's best friend. He is also extremely stupid, as will be evidenced later. Squidward works as the cashier at the Krusty Krab and despises his job. He is annoyed by SpongeBob and Patrick, but would secretly miss them if they were gone. Gary is a snail who is SpongeBob's pet, and has a cute meow. The final major character is Sandy, a squirrel from Texas who lives in a biosphere under the sea. When she leaves the biosphere, she wears an astronaut-like suit so that she can breathe. When Patrick or SpongeBob enter Sandy's biosphere, they wear bowls of water over their heads.

As evidenced by Sandy's character, the show prides itself on surrealism. It takes events, objects or characters that do not belong in an underwater environment and places them in Bikini Bottom. For example, the characters often go to the (underwater?) beach to lie by the shore. Balloons are filled with air. Campfires burn until the characters remember that they cannot have campfires underwater. The flames then magically go out, illustrating the reflexivity of the show's animated form.

The characters are also highly anthropomorphized, as is the case in almost all animal-based animation. The fish walk on their fins. Squidward, with his six tentacles, stands upright with two tentacles for arms. The other four tentacles move in pairs of two to make him bipedal. The plotlines themselves portray human interests, such as whether SpongeBob is tough enough to get into certain clubs and whether he can pass his driving test. Steve Baker criticizes studies of "talking animal stories" for placing too much focus on the human-oriented narrative. Often, he argues, scholars mention at the beginning of a study that the characters are animals, but then immediately turn to the human interests in the narrative. According to Baker, the visual representation of characters as animals serves to decenter the human narrative (131). Whereas the narrative may illustrate human concerns, there is no way for the viewer to escape the fact that the characters are animals due to the animal imagery. The visual and verbal representations are intertwined, and cannot easily be separated. Rather than the story consisting of a visual narrative of animals and a verbal narrative of humans, a more complex story of animals pursuing human interests emerges. Human interests and choices become applicable to animals in a subtle way, making it easier for humans to make decisions for animals.

Neoteny is one of the elements of the visual design of anthropomorphized animals that permits human control of the image. Neoteny renders a person, object or image "cute" through infantile physical features, such as the large eyes of a baby, which often elicit the urge to protect and nurture. All of the characters in *SpongeBob Squarepants* have large eyes, as occur in infants. In addition, Patrick, SpongeBob and Mr. Krabs all have large fat torsos with stubby appendages, much like babies. While animation commonly uses the trope of neoteny to make characters appear cute, lovable and purchasable, it is especially problematic in animal characters.

Neoteny helps naturalize the unequal power relations between animals and humans by visually depicting animals as creatures needing human nurturing and care. By depicting animals as infantile (even animals who within the narrative of the show are portrayed as adult humans), the connection can be made that animals, like infants, need adult humans to make decisions for them. For example, it may be in the best interest of an animal to be confined in a cage in a zoo. Similarly, people confine their pets to their houses or yards, claiming it is in the best interest of the pet, when many stray animals may live worthwhile lives. This is why the animal defense movement uses the term "companion animal" instead of "pet." "Companion animal" denotes an animal allowed to pursue its own interests that chooses to live with humans as a companion. Neoteny of animal images contributes to the public perception that animals are like infants and need humans to make decisions for them.

SpongeBob represents a sentient animal that is highly disregarded in society: a sponge. Today most sponges are synthetic and not made from animal bodies (potters' sponges are an exception). However, the shape of SpongeBob portrays a sponge that is used for work in the kitchen, rather than a natural sponge that would grow in the ocean. Sponges harvested from the ocean are often cut into these shapes. Indeed, SpongeBob loves to work in the kitchen at the Krusty Krab. He genuinely enjoys his work. Moreover, he uses his body to absorb fluids as part of the surrealism of the underwater show. In an episode where he takes Mr. Krabs's daughter to the prom, SpongeBob knocks over the punch bowl. He then proceeds to use his body to absorb the punch and squeezes himself out into people's glasses. In the episode "Gary Takes a Bath," SpongeBob is frustrated by Gary's attempts to avoid bathing. Finally, SpongeBob absorbs all the water in the bathtub into his body, and walks around shooting the water out of himself in an effort to hit Gary. Ultimately, he fails, falls in a mud puddle and needs a bath himself. This plotline shows the affection with which pets should be treated. It is important to take care of pets, bathe them, feed them and buy them toys as SpongeBob often does for Gary. Pets are articulated as a necessary place for people to purchase commodities for their care, thus advancing capitalism.

In addition to the instances of using animal bodies for work, catching jellyfish is a major activity for Patrick and SpongeBob. They go to Jellyfish Fields and catch a jellyfish with a net to "milk" for jam on their sandwiches. The tentacles of the jellyfish are visually transformed into udders for the process, alluding to the milking of cows and masking the conditions in which dairy cows live. The jellyfish roam in open fields, in contrast to the living conditions of most dairy cattle. These instances illustrate a subtle cultural depiction that animals enjoy, or at least do not mind, using their bodies for labor.

Also, the bodies of the animals of Bikini Bottom do not become injured. SpongeBob is often diced into many pieces, such as while jumping rope or being dragged through cheese graters while hanging behind a boat. His arms and legs often come off, but he grows them back immediately. Patrick often injures himself as well, but does not appear to know that injuries are supposed to hurt. Patrick is so unintelligent that he cannot work or even recognize his own parents. Interestingly, a fifty-two-year-old fan states regarding *SpongeBob Squarepants*, "Besides silly jokes, I like that it really doesn't have violence in it and nobody really gets hurt" (Johnson D1). According to this fan, the dismemberment of animal bodies is not violence. The lack of consequences of the violence may actually be a selling point of the show.

In the episode "Hooky," Mr. Krabs warns SpongeBob that the hooks are back. He tells SpongeBob that SpongeBob should never go near the hooks; the fishermen would drag him above the water and either sell him in gift shops or pack him in tuna cans. Patrick takes SpongeBob to the "carnival," but it is actually the hooks. Although SpongeBob urges him to stop, Patrick continues eating the cheese off the hooks and insists there is nothing to worry about. Finally, a hook rushes upward, carrying Patrick with it. This is interjected with quick cuts of live-action images of fishermen in a boat reeling in a line. The world above the ocean, the realm of humans, is always portrayed in live-action images. SpongeBob cries for his dear friend, who eventually floats down. Patrick claims he jumped off and had the ride of his life. Clearly, smarter animals know enough to stay away from the hooks. A fisherman catching Patrick is a result of his not being smart enough, the incident thus propagating the idea that the animals killed for food are "dumb" and therefore deserve the treatment. They begin riding the hooks until Mr. Krabs catches and admonishes them. He asks, "Do you know what happens when you don't float back down?"

"Gift shops?" asks SpongeBob.

"Worse! You end up water-packed in a can of tuna with nothing to look forward to but the smell of mayonnaise." Again, the idea that an animal could look forward to anything while packed in a tuna can masks the idea that humans actively kill animals and package their corpses for food. Somehow the body in the tuna can is unharmed. Other animals such as dolphins and sea turtles are often caught in industrial tuna nets as well. The fishermen appear to be recreational fishermen, not industrial; thus the show shies away from critiques of the tuna industry.

Finally, SpongeBob catches his clothing on a hook and someone attempts to reel him in. In a flash of the images in SpongeBob's mind, the audience sees his sponge body smushed into a tuna can with his face staring out. Mr. Krabs convinces him to take off his clothes to be saved, despite the fact that girls are watching in the restaurant. Rather than the hook tearing into his body like the actual hooks of fishermen, this hook is only caught on his clothes. His body is often injured in the show, so the only reason for not injuring it in this case appears to be an attempt to cover up the ways that fishing practices damage animal bodies. In the end, Patrick, the stupidest animal, is caught by the fishermen and packaged in a tuna can. A delivery truck brings Patrick, inside a can, to his house. In the last line of the episode, Patrick, unscathed and in one piece in the tuna can, yells, "Help! Somebody get a can opener!" Ultimately, animals are unharmed and good sports about humans' practices of using animals for food and labor. Episodes like this one perpetuate the ways that urban dwellers overlook and compartmentalize the realities of suffering in food animals. Children are able to maintain the fantasy that food animals are not harmed. For adults, the show allows the continuance of the dichotomy between those animals which are loved, like pets, and those which are overlooked, eaten and worn.

CONCLUSIONS

Animal representations are one of the major ways in which people in the United States, particularly urban people, interact with animals. Visual communication about animals through animation, wildlife documentaries and other forms of media creates and reinforces ideologies about animals. Our other discussions about animals spring from these places, as these are our learning tools. Children grow up with animated characters and are taught implicit messages about their relationships with animals through SpongeBob and others. The ideology that is learned about animals through representations is still heavily ingrained in American society.

These representations serve as a teaching tool to show people how to differentiate animals into groups such as pets or products. These distinctions are held to, as is evidenced by the cultural disdain for eating dogs in the United States. Dogs hold the space of the "good animal" within society, one that is okay to love and lavish with gifts. Animals are also considered workers for capitalistic purposes. Through representations, worker animals, when shown at all, are usually shown enjoying the work while depicted in conditions unlike those of actual worker animals. More troubling, animated animals sometimes participate actively in their own consumption as a food source by others. *SpongeBob Squarepants* serves as an illustration of one instance of this in society, but represents a larger trend toward rendering animal suffering invisible in order to preserve the structure of late consumer capitalism.

Animation is a key site of animal representations. Much of animation, and clearly *SpongeBob Squarepants*, is intended for humorous purposes, but this does not negate the potential consequences of such representations for living animals through insidious speciesism. Seen through the giant eye-balls of SpongeBob Squarepants, certain marine lessons are highlighted for viewers, such as that recreational fishing (both hooking and being hooked) is fun, that marine animals need human nurturing and that marine animals have interests and needs similar to those of humans (which is unlikely). The fact that human practices are at the heart of most threats to marine animals is ignored. Industrial fishing, pollution, offshore drilling and climate change are not present in the narrative. Basic awareness of these issues can be included within children's programming in a fun way that could encourage children to make small changes, such as recycling or changing their eating habits. However, this would fly in the face of major advertisers such as oil and gas companies, fast food restaurants, fish producers, car manufacturers and many others. Advanced capitalism creates problems for wildlife, such as deforestation, and for animals used as food, including skin lesions and pneumonia in pigs as a consequence of being urinated on and breathing in urine in extremely overcrowded conditions. Such conditions perpetuate speciesism in the interest of perpetuating capitalism.

NOTES

1. I argue that a beast is one who does not comply with white male notions of civility and rationality.
2. While well-intentioned and perhaps necessary for the advancement of women at the time, this move served to reify patriarchal standards for intelligence instead of critiquing those very criteria by which women were judged to be beasts.

3. The abundance of nonhuman animal characters in U.S. animation is so great that I cannot cite them all here. I draw from Jeff Lenburg's *The Encyclopedia of Animated Cartoons*, Giannalberto Bendazzi's *Cartoons: One Hundred Years of Cinema Animation* and Leonard Maltin's *Of Mice and Magic: A History of American Animated Cartoons*. More information about the examples I use can be found in any of these books.

REFERENCES

Adams, Carol J. *Neither Man nor Beast: Feminism and the Defense of Animals.* New York: Continuum, 1995. Print.
———. *The Sexual Politics of Meat: A Feminist-Vegetarian Critical Theory.* New York: Continuum, 1990. Print.
Baker, Steve. *Picturing the Beast: Animals, Identity, and Representation.* Urbana: U of Illinois P, 2001. Print.
Bendazzi, Giannalberto. *Cartoons: One Hundred Years of Cinema Animation.* Bloomington, IN: Indiana UP, 1994. Print.
Berger, John. "Why Look at Animals?" *The Animals Reader: The Essential Classic and Contemporary Writings.* Ed. Linda Kalof and Amy Fitzgerald. Oxford: Berg, 2007. 251–261. Print.
Burt, Jonathan. "The Illumination of the Animal Kingdom: The Role of Light and Electricity in Animal Representation." *Society & Animals* 9 (2001): 203–228. Print.
Chozick, Amy. "Return to the Big Screen for Spongebob." *New York Times* 5 Mar. 2012, late ed., sec. B: 4. Print.
DeLuca, Kevin Michael. *Image Politics: The New Rhetoric of Environmental Activism.* New York: Guilford Press, 1999. Print.
Hendershot, Heather. *Saturday Morning Censors: Television Regulation before the V-Chip.* Durham, NC: Duke UP, 1998. Print.
Johnson, L. A. "Soaking Up Viewers: Adults as Well as Children Tune In to the Wet and Wisdom of *Spongebob Squarepants.*" *Post-Gazette* 2 Jul. 2002, sooner ed., sec. arts & entertainment: D1. Print.
Keveney, Bill. "Nickelodeon Comes of Age." *USA Today* 12 Jan. 2001, final ed., sec. life: E1. Print.
Lenburg, Jeff. *The Encyclopedia of Animated Cartoons.* 2nd ed. New York: Checkmark, 1999. Print.
Major, Rose. "Turner Seeks Quick Return from Grown-up Cartoon." *New Media Markets* 27 (Sept. 2002): Electronic Version.
Mitchell, W. J. T. *Picture Theory.* Chicago: U of Chicago P, 1994. Print.
Rosenthal, Phil. "Is Spongebob Close to Being Washed Up?" *Chicago Sun-Times* 13 May 2002, sec. features: 45. Print.
Singer, Peter. *Animal Liberation.* 2nd ed. New York: Random House, 1990. Print.
———. "Animal Liberation or Animal Rights?" *The Animals Reader: The Essential Classic and Contemporary Writings.* Ed. Linda Kalof and Amy Fitzgerald. Oxford: Berg, 2007. 14–29. Print.
Stabile, Carol A., and Mark Harrison. "Prime Time Animation: An Overview." *Prime Time Animation: Television Animation and American Culture.* Ed. Carol A. Stabile and Mark Harrison. London: Routledge, 2003. 1–11. Print.
Wasko, Janet. *Understanding Disney: Manufacture of Fantasy.* Cambridge, UK: Polity, 2001. Print.
Werts, Diane. "Toon Town after Hours: Cartoon Network Wants Baby Boomers to Stay Tuned to Its Expanded Block of Shows for Adults." *Newsday* 12 Jan. 2003, sec. fanfare: D10. Print.

Part II
Implication

6 Stepping Up to the Veggie Plate
Framing Veganism as Living Your Values

Carrie Packwood Freeman

What does meat say about meat eaters? As an animal activist, I often publicly opined on what was *wrong* with meat and other animal products—the suffering, the unnecessary killing, the pollution, the wilderness destruction, the clogged arteries, etc. But another way of approaching the issue, and a way of doing so more positively, is to ask what is *right* about choosing to eat plant-based foods. What does a plant-based diet say about plant eaters?

As a communications scholar, I decided to study how animal rights organizations attempt to entice meat-eating Americans into choosing veganism. And although that involves highlighting problems with animal flesh, eggs and dairy, it also involves constructing solutions with respect to eating solely plant-based foods and motivating people to see themselves as vegans—with veganism, perhaps surprisingly, being constructed as a natural fit for their values. In this chapter I share the sixteen main values representing the presumed identity of a vegetarian in the "go veg" campaigns of five major U.S. animal rights organizations. I evaluate how this construction does serve and could better serve the motivation and identity function of the social movement framing process.

VEGETARIANISM IN THE UNITED STATES

For over a decade, most major animal rights organizations (AROs), and even some animal welfare groups, have made farmed animals a primary focus, realizing "food animals" comprise the vast majority of nonhuman animals (NHAs) killed in the United States—over ten billion land animals, mainly birds, and at least as many sea animals annually (Singer and Mason 112). And the number of animals killed annually continues to slowly rise (FARM).

Donna Maurer has examined the history of vegetarianism as a movement in the U.S., finding that vegetarianism peaked in the mid-1800s and again in the 1960s and 1970s (3). Ever since, vegetarianism has held a small but steady contingency without growing significantly. In spite of the animal rights and vegetarian movements, Americans' per capita consumption of meat kept increasing in the 1980s and 90s (Maurer 15) but has finally just started decreasing in recent years ("Peak Meat").While many Americans say they are more frequently choosing vegetarian meals, only about 3% of the population is fully vegetarian, eating no animal flesh. At least 1% of these, or approximately a million people, are vegan and eat no animal products whatsoever (Maurer 16–17; Singer and Mason 4–5, 187). The typical person attracted to vegetarianism is a young, white, middle-class, atheist female (Maurer 8).

Margaret Visser, a cultural historian, suggests that "vegetarianism can be viewed as a modern response to dealing with the endless choices engendered by a consumer society that discourages the appearance of overconsumption" (qtd. in Maurer 138). And while environmental sustainability is a motivation for some vegetarians, the main reasons people *say* they go vegetarian is for health and/or ethics. People who go vegetarian for ethical reasons tend to be more committed to remaining vegetarian. So, Maurer posits, "promoting concern for animals and the environment is essential to the advancement of the vegetarian movement" (45) because health-motivated vegetarians may be tempted by the convenience of a meat-based diet and new lower-fat meat items (126).

COMMUNICATION TACTICS OF U.S. VEGETARIAN ADVOCATES

The vegetarian movement's ideology is based on three core tenets: vegetarianism supports human health, compassion for NHAs and environmental sustainability (Maurer 71). While advocacy organizations tend to agree on the merits of these tenets, they sometimes disagree on how to market them. For example, their advocacy materials may choose to promote one benefit over others, or they may shy away from the word *vegan* as it is less familiar and may seem extreme to the general public (95). Some even opt to replace the familiar but culturally loaded term *vegetarian* with the more benign and clinical term *plant-based diet* (96).

The main framing debate within vegetarian advocacy is whether to promote altruistic ethical benefits and a collective identity or to promote individual, human-health benefits (Maurer 119). Maurer has found that, for wider appeal, vegetarian campaigns often chose to emphasize health. However, a campaign promoting a strong vegetarian identity based on ethical principles, although it attracts fewer people, can inspire a stronger commitment than a more vague and mainstream health appeal. Yet a quarter of Americans say they are actively reducing their meat consumption, primarily based on self-interest, such as health, rather than on animal or environmental protection ("Advocating Meat"). While consumers view vegetarian foods as healthy, they also generally believe that some animal products, such as dairy, chicken and fish, are also healthy.

Vegetarian advocates are challenged by survey findings that reveal 80% of Americans do not intend to ever fully eliminate meat from their diet, based on concerns that it may be unhealthy to do so and their overall preference for the taste of meat. Therefore, survey data reveals that it would be more effective for vegetarian advocates to promote meat *reduction*, rather than vegetarianism, and that reduction might lead to vegetarianism down the line ("Advocating Meat"). Whereas the health rationale in particular, and to some degree the environmental rationale, is more useful at encouraging people to *reduce* meat consumption, the animal suffering rationale is most effective at motivating people to *eliminate* meat ("Advocating Meat").

Maurer warns that if vegetarianism becomes just another healthy lifestyle consumer choice, it risks being reductively perceived as a form of "feminine asceticism" (145) and loses its ideological edge as a "public moral good" (126). Perhaps it is advantageous that a significant portion of the vegetarian movement is comprised of animal rights organizations whose campaigns tend to promote more ethical urgency and inspiration than do the campaigns of solely vegetarian organizations (68–69).

COMMUNICATION STRATEGIES OF SOCIAL MOVEMENT ORGANIZATIONS

Social movement organizations (SMOs), like those in animal rights, struggle to rhetorically transform a hegemonic view of reality in the dominant discourse, in this case an instrumental view of certain NHAs as food objects. SMOs need to convince the public not only that the commonly accepted view of reality is based on a faulty premise, but that the situation deserves to be defined as a problem that warrants their immediate attention (Stewart, Smith, and Denton 52).

In the quest to create meanings in support of their worldview, SMO communicators can find guidance in sociological literature on the framing process (Johnston and Noakes 1–2). Framing is based on "principles of selection, emphasis, and presentation composed of little tacit theories about what exists, what happens, and what matters" (Gitlin 6). SMOs must construct collective action frames that define problems and culprits; demonstrate the problem's severity and urgency; suggest logical, realistic solutions; and encourage participation based on shared identity and values. William Gamson defined the three components of collective action frames as: agency (we can fix it if we work together), identity (side with us) and injustice (7). David Snow and Robert Benford defined the three core tasks of framing more generally as: diagnostic, prognostic and motivational (464).

I am interested here in the *motivation* and *identity* components of collective action frames, as SMOs must construct a compelling incentive that serves as meaningful inspiration for people to engage in collective action in view of the proposed solution (Benford and Snow 614). To garner this support, motivational frames often rely upon an appeal to shared values, demonstrating alignment between the goals of the SMO and those of the target audience. Snow, Rochford, Worden and Benford identify four main frame alignment processes from which SMOs could choose: bridging ideas to reach sympathetic but unmobilized supporters; amplifying important beliefs and values to demonstrate their relevance to the SMO's issue; extending the issue's relevance to other related social issues; and transforming people's views on the issue to cast it in a new light (467).

To increase the resonance of frames, AROs should seek credibility by using arguments that are authentic to their beliefs, truthful and logically consistent. To be resonant, frames should also create salience by appealing to key, broad, culturally accepted values and connecting them to the audience member's personal everyday life (Benford and Snow 619–621; Tarrow 118). Additionally, an ARO should promote a clear set of simple values, more so than facts, to accurately reflect its principles and promote a moral vision (Lakoff 74)—in this case, a primarily vegan society that does not domesticate and exploit fellow animals.

METHOD

Grounded by social movement framing literature, I ask: To which values are AROs appealing? And how are AROs creating alignment between their values and those of the public? To examine "food animal" advocacy, I follow Stuart Hall's cultural studies textual analysis method, examining words and images in context to uncover the themes and assumptions grounding the construction of ideas. The food-advocacy text sampled includes vegetarian or vegan materials as well as anything addressing the human practice of farming animals or fishing them to use for food. I include both electronic and print materials being used by five selected AROs as of January 2008. Electronic materials include web pages and self-produced video footage (including advertisements and animal cruelty footage). Print materials include vegetarian starter guides, pamphlets, advertisements and collateral pieces such as stickers, clothing, buttons and posters.

To be comparable and relevant, AROs selected for this study had: an animal rights mission supporting veganism in contrast to a more moderate welfare mission promoting "humane" meats; campaigns providing a variety of print and electronic advocacy pieces aimed at the public; and a national presence within the United States. The following five organizations, listed from largest to smallest, most fully met criteria for inclusion: People for the Ethical Treatment of Animals (PETA), Farm Sanctuary (FS), Farm Animal Rights Movement (FARM), Compassion over Killing (COK) and Vegan Outreach (VO).

PETA was founded over 25 years ago in Washington, D.C. Now head-quartered in Norfolk, Virginia, PETA has expanded to become the largest animal rights group in the world, with more than 150 full-time paid staff, international offices and more than 1.8 million members and supporters ("About PETA"). FS was founded in 1986 and has grown into the largest farmed animal rescue organization in the nation, operating sanctuaries in New York State and California, with more than 100,000 members and 75 paid staff ("About FS"; "FS Financial"). FARM, located in Bethesda, Maryland, with 7 paid staff, claims to be the oldest animal rights group dedicated to farmed animal issues; it started in the early 1970s ("About FARM"). The youngest group, COK, was founded in 1995 as a high school group in Washington, D.C., and now has 6 paid, full-time staff ("About COK"). VO is a highly focused group which began in 1993; it primarily operates from Tucson, Arizona, with just 3 paid staff members and a host of volunteers to hand out its pamphlets on college campuses ("About VO").

While marginalized in the public sphere, the advocacy messages of these AROs have likely reached hundreds of millions of Americans in an attempt to make production and consumption of animal products an ethical issue. It represents an important challenge not only to mainstream food industry discourse but also to American social norms and basic ideals concerning whom it is morally acceptable to use and kill and who pays the cost for America's food choices.

FINDINGS

To demonstrate veganism's fit with consumer identities, AROs most promi-nently appeal to our presumed values of compassion, respect for animals as subjects (not objects), healthfulness, environmentalism and moral consis-tency. They also appeal to other values such as: desire to improve the world and make a difference, choice, pleasurable and convenient food, belong-ing, life, concern for fellow human beings, honesty, populism, naturalness, freedom and American pride. I categorize and discuss these values in three major categories: altruistic, idealistic and personal. I'll share a few exam-ples of how AROs emphasize and frame each value.

Altruistic Values

Compassion and Caring for NHA Suffering and Aversion to Cruelty.

AROs spend a lot of time educating consumers about factory farm and slaughterhouse cruelty, implying consumers are compassionate and caring enough to be offended, an assumption which is also explicitly declared. For example, FS's new slogan is "a compassionate world starts with you." PETA's *Meet Your Meat* video ends with celebrity vegetarian Alec Baldwin telling viewers to think about the cruelty they have seen, to choose "compassion" and to go vegetarian as "millions of compassionate people" have decided to do. VO's *Why Vegan?* booklet declares "we can choose to act with compassion by boycotting animal agriculture. Making humane choices is the ultimate affirmation of our humanity." COK's *Vegetarian Starter Guide* creates a good versus bad dichotomy under the title "Choosing Compassion over Killing" by asking "Do we want to support kindness and mercy, or do we want to support cruelty and misery?" and requests that readers "take a stand for compassionate living." FARM literature often describes people as "caring" and refers to those who speak out for farm animals on World Farm Animal Day as "people of conscience."

Respect for the Sentience and Individuality of Other Animal Subjects

Built into the assumption that someone is compassionate toward NHAs is the idea that the person respects the other animals' ability to feel and does not want them to suffer. The concept of sentience, as I use it here, involves not only experiencing pain, but also experiencing emotions, thought or consciousness. AROs typically do not use the word "sentience"; rather they tend to say that NHAs "feel" or "suffer." To refer to NHAs as subjects, not objects, all AROs use gendered or personal pronouns like *he, she* or *I* when referring to farmed animals instead of following the common American practice of calling an animal *it*.

All AROs include frequent messages to ensure the public that farmed animals are sentient, often comparing their capabilities to those of cats and dogs or sometimes to other animals including, less frequently, humans. An example of a pet comparison is FARM's vegetarian postcard which states, "animals raised for food are just as intelligent, lovable, and sensitive as the animals we call *pets*." An example of a human comparison is PETA's teen vegetarian booklet, which declares, "animals are like *us*" and proceeds to describe farmed animals doing what would normally seem like human activities, such as pigs playing video games, turkeys playing ball, cows babysitting and fish gardening.

FS's "Sentient Beings" campaign seeks elevated legal status for U.S. farmed animals; that is, it advocates that they be classified as sentient beings as they are in Europe. The leaflet for the campaign is titled "Farm Animals Have Feelings Too" and says these animals are "sentient beings—capable of awareness, feeling and suffering" who "deserve to be treated with respect." This is contrasted with pictures of farmed animals in extreme confinement and quotes from industry that compare them to machines and manure, a tactic used by many AROs to demonstrate how industry commodifies animals.

To showcase farmed animals as individuals, all AROs portray pictures of them making direct eye contact with the reader. Several vegetarian starter guides describe the personalities of each rescued animal and display his or her portrait and individual name, such as Travolta the cow, Emery the chicken and Ashley the turkey. The descriptions reveal personality traits, such as friendliness, talkativeness, playfulness and preferences for certain foods such as apples or green grapes. In FS's *Guide to Veg Living*, a photo shows a goose, Bing, happily spreading his wings in a pond and honking with gusto, and another photo shows a piglet, Rudy, standing proudly and defiantly in the grass with the low camera angle putting the viewer in the position of looking up at him so that he appears larger than life.

Several PETA publications feature an "Amazing Animals" section praising animal abilities. Here are examples of PETA's opening sentence descriptions for each species:

> Chickens are inquisitive, interesting animals who are thought to be as intelligent as cats, dogs, and even some primates; Pigs are curious and insightful animals thought to have intelligence beyond that of an average 3-year-old human child; Fish are smart, sensitive animals with their own unique personalities; Cows are intelligent, loyal animals who enjoy solving problems; Turkeys are social, playful birds who enjoy the company of others; Geese are very loyal to their families and very protective of their partners and offspring.

PETA dedicates more space to fish than does any other ARO, and it is the only group that talks about fish sentience in terms of intelligence and personality. In general, mammals such as cows and pigs seem to be the most popular animals for all AROs to display, with birds being the next most popular.

Some messages overtly request that consumers view farmed animals as more than food objects. The very title of PETA's popular video, *Meet Your Meat,* conveys the incongruous idea that consumers can see farmed animals both as individual subjects while alive and as objects after death. COK's print advertisement displays a cow's face reflected in a woman's eye and asks teen girls to "see her as more than a meal." FS's print ad features a young pig, Truffles, who challenges the viewers to "look me in the eyes and tell me I'm tasty," and a sticker showing an illustration of a chicken stating, "I am not your breakfast, lunch or dinner." Similarly, PETA has several collateral materials with an illustration of a chick declaring, "I am not a nugget" and telling viewers that pigs and fish are "friends not food." Emphasizing friendship, as several AROs do, challenges the solely instrumental relationship we typically have with farmed animals.

Concern for Fellow Human Beings

While AROs emphasize care for nonhumans, some also show how veganism is compassionate toward humans, especially innocent people who are wronged by meat-industry practices. This anthropocentric altruism is particularly emphasized by FARM, PETA and FS, all of which have antifactory farming campaigns to fight human hunger, worker exploitation or the polluting and health contamination of rural neighborhoods. For example, PETA warns that "profits are put before people" by government and factory farmers, so readers are encouraged to go vegan to "stop these exploitative industries and promote a world of compassion." FARM is the only ARO that has an antihunger campaign; this campaign, called Well-Fed World, promotes "plant-based diets" to reverse starvation rates as the worldwide consumption of unsustainable animal products and factory farming result in resource depletion and unequal distribution of food.

Environmentalism

Of increasing popularity is an appeal to people's concerns for how our food choices affect nature (and its human and nonhuman inhabitants). PETA, FS and FARM have print and online pieces specifically dedicated to framing animal agribusiness as environmentally destructive, commonly featuring photos of pipes spewing manure into cesspools next to factory farms. PETA's *Chop Chop* leaflet claims one can't be a "meat-eating environmentalist" and visually equates a pork "chop" to trees being "chopped," providing details on meat's association with global warming, pollution, excessive resource use and damage to oceanic life. The three print pieces of FS's "Veg for Life" series all mention environmental degradation, using verbs such as *eroded, ruined, contaminated, compromised, mismanaged and ransacked* and declaring that the number two reason to go vegetarian is because "much of our water and fossil fuel supply is squandered for livestock rearing." FS's gray brochure titled "Factory Farming: Destroying the Environment" emphasizes the pollution of nature and our bodies by showing photos of cesspools, chemical plants, pharmaceuticals and a fish kill. FARM's "Bite Global Warming" campaign, found at coolyourdiet.com, is built around a 2006 report of the United Nations Food and Agriculture Organization that lists animal agriculture as an even bigger "culprit" in greenhouse gas emissions worldwide than the transportation industry, a climate change fact increasingly cited in many other ARO messages.

Life

To show the value of the life of all living beings, AROs sometimes specifically problematize the killing and death of NHAs, not just their suffering. To promote saving life, COK names its main veg starter guide sections "Saving Ourselves," "Saving Animals" and "Saving the Earth." And consider the use of the word *life* in FARM's "choose life" slogan for its Great American Meatout and in its Thanksgiving campaign declaring that killing innocent animals "betrays the life-affirming spirit" of the holiday, asking viewers to "celebrate life." In a more direct life-saving appeal, FS has stickers showing cows and chickens with a statement reading, "She wants to live and her life depends on YOU!" AROs often talk about the number of animal lives saved by vegetarians per year, or conversely, how many lives meat eaters take.

Human lives can also be saved via veganism. In FS's "Veg for Life" vegetarian campaign, one can interpret the word *life* to mean that a healthy vegetarian diet can save one's own life and/or the lives of farmed animals. For example, FS's and PETA's main vegetarian guides title the recipe section "Recipes for Life" to indicate food choices that result in healthy living bodies. Life could also connote time, suggesting people should eat vegetarian for the rest of their lives.

Freedom

Freedom is a key human rights value in America, so AROs attempt to apply it to *all* animals having the right to freedom over their own lives and bodies. PETA's *Chew on This* DVD declares that "everyone wants to be free," meaning NHAs too. But, this example to the contrary, AROs make few direct references to *freedom*; rather, it is implied. AROs' consistent emphasis on animals' extreme confinement and immobility in factory farms implies that Americans will find high levels of restriction to be unfair.

Although infrequent, direct analogies to human confinement are sometimes made, such as COK's pork leaflet comparing the confinement of pigs "stuck" in gestation crates to how frustrated and uncomfortable people would feel being "stuck" in a car in traffic for years. It says pigs are "unable to move freely" and "can't even walk or turn around," as they are in a pregnancy cycle in which they are moved continuously between gestation and farrowing crates. The word *cycle* is used by many AROs to describe the treadmill of reimpregnation faced by pigs and cows to maximize industry profit.

Freedom is associated with wide-open spaces, as in the American West, which may be why FS emphasizes space on its new home page, with lots of white space, a picture of rescued animals enjoying the sun and a strip of grass across the bottom. Blue sky, sun and grass are often represented in ARO pictures of contented animals to emphasize their relative freedom (presumably in a sanctuary) in contrast to the darkness, filth and discomfort of captivity on factory farms.

Desire to Improve the World and Make a Difference

ARO messages indicate that people seek meaning in their lives by improving the world and making a difference. COK encourages readers to "Make a difference. Start today!" and the back of its vegetarian starter guide states in bold that "every time we sit down to eat, we can make the world a better place." Similarly, VO's *Even if You Like Meat* booklet tells readers "every time you choose compassion, you're making a difference." To emphasize personal empowerment, FARM uses the slogan "Stop global warming one bite at a time," describing the "power" of our food choices and how they "matter." Similarly, FS has a radio PSA for Earth Day that says the "power is on your plate" to protect the earth "every time we eat."

Note that in many of these slogans there is a time element emphasizing the ease with which a person can make a difference through vegetarianism *every day* because it allows him or her to improve the world "at every meal" or "one bite at a time." These AROs transform the mundane act of eating into a convenient form of activism for those who do not necessarily want to dedicate time to being a traditional activist or do not have the money to donate to causes.

Further indicating the importance of a switch to vegetarianism, AROs often claim that vegetarianism is the *best* way to help animals and the planet. FS's *Veg for Life* brochure says "eliminating ALL animal foods from our diets is the single most important step we can take to be kinder to animals, ourselves and the Earth." And PETA often cites a statement of vegetarian Sir Paul McCartney: "If anyone wants to save the planet, all they have to do is just stop eating meat. That's the single most important thing you can do."

Making a difference is also connected with self-interest in feeling good about oneself. FS says that through veganism you will "feel good because you make the world better," stating that vegetarians enjoy better "mental health and feel good knowing they are working toward improved health and well-being for themselves, animals and the environment." Regarding mental health, veganism is often framed as a personal growth goal. COK's veg guide section on transitioning to a vegetarian diet reads like a life coach's plan praising new vegetarians because they have "made it!" and deserve a "pat on the back!"

Idealistic Values

Naturalness

Naturalness is a value the public relates to food's healthfulness, in contrast to artificial foods. PETA and COK frame cow's milk as natural for calves to drink but unnatural for humans, while meat was only once (by PETA) framed as unnatural for humans to eat. However, in this section I focus less on the health angle of the naturalness value and more on how AROs suggest that what is natural for animals and what is more traditional for agriculture is preferred to what is artificial or industrialized, such as the genetic modification of animals and large-scale factory farming.

Mechanization, metal and concrete abound in images of animals confined in warehouses, left to die in garbage cans and disassembled on the slaughterhouse assembly line. The feel is cold, dark, gray, dirty and industrial. This unnatural environment is juxtaposed against the cleanliness and brightness of portraits of animals, presumably in a sanctuary or smaller farm, surrounded by natural elements of sun, grass, hay, wooden fences and ponds. To a lesser degree, wild animals, particularly fish and sometimes turkeys, are shown living in nature.

AROs often directly refer to the unnaturalness of the practices, conditions and animals' bodies in modern animal agribusiness. For example, FS says the number ten reason to go vegetarian is that "farm animals are usually prevented from engaging in instinctual behavior and live a fraction of their natural lives." And VO's booklets cite Michael Pollan saying of a battery caged hen, "every natural instinct of this animal is thwarted." These stifling factory conditions are contrasted with the descriptions of how these species would behave in nature.

The industry demands animals grow to an unnatural weight, the cruelty and artificiality of which is highlighted by AROs. FS's video on the turkey industry explains how farmers alter the shape of the birds to meet consumer demands for turkey breasts, stating that this "anatomical manipulation" has made male turkeys so large that it is impossible for them to "mount and reproduce naturally," so female turkeys must be "artificially inseminated." And COK's veg guide explains how birds "grow so abnormally fast due to selective breeding and growth-promoting antibiotics" that they suffer organ failure and lameness, collapsing under their bulk. Many AROs critique the unnatural diet and medications that agribusiness uses to fatten animals, likely inciting our fears over frankenfoods and superbugs. Chickens are "dosed with a steady stream of drugs" (PETA) and cows are "fattened on an unnatural diet of grains and 'fillers' (including sawdust and chicken manure)" (COK). FS identifies bovine growth hormone as the reason cows "produce ten times more milk than they would in nature," showing engorged udders practically dragging on the ground.

ARO environmental messages claim such out-of-sync agribusiness practices contaminate the purity of nature. FS's factory farming brochure cites a statement of the Worldwatch Institute: "overgrown and resource intensive, animal agriculture is out of alignment with the Earth's ecosystems." To further indicate the artificial, the brochure shows photos of medication as well as fumes coming from an agricultural chemical plant. Related to this, FS's, COK's and PETA's vegetarian guides all mention contamination in the resulting animal products humans eat, saying that animal products are a health risk because they contain unnatural ingredients like pesticides, drugs and other chemicals.

Honesty

Consumers need truthful information to make informed decisions, and AROs blame agribusiness for misleading consumers and hiding the ugly truth of factory farming; AROs see it as their job to give people a reality check. COK places primary emphasis on the honesty value with its campaign against fraudulent "humane" farming labels and its television ads showing flabbergasted consumers being served a rare "side of truth" at a fast-food restaurant.

In an interesting twist on honesty, PETA's *Chew on This* DVD accuses *parents* of being dishonest to children when the narrator says "you shouldn't have to lie to your kids" about where their food comes from. This assumes that adults know that the reality of farm animal suffering and death is gruesome enough to upset the kids and possibly keep them from eating meat.

American Populism, and Big Business and Government Responsibility.

This broad category overlaps with the values of naturalness, honesty and concern for human well-being, as ARO messages capitalize on an assumed public mistrust of the exploitative and irresponsible tendencies of big business and, in some cases, government. This idea of American populism suggests that AROs assume people want corporations and the political elite to be held accountable in cases where they take advantage of the innocent little guy. For example, all AROs critique modern agricultural practices specifically on the basis that they are contemptible as "factory farming," "corporate agribusiness" or an "exploitative industry," in contrast with the bucolic values that consumers may have for wholesome traditional or family farming, considered a responsible business of everyday hardworking people. Hence, AROs rhetorically center blame on agri*business* more than agri*culture*. AROs generally do not insinuate that small or "family farms" are nearly as problematic. They blame factory farming, in particular, for why cruelty is standard, food is unwholesome, the earth is polluted, workers are exploited and consumers are misled.

PETA and FS appeal to these populist values the most, as they both have online sections discussing the exploitation of workers and the contamination of rural communities by animal agribusiness. The implication is that industry is greedy and callous and fails to demonstrate justice, respect, responsibility and decency toward the common man. The jobs agribusiness provides are described as dangerous, dirty and low paying. PETA cites workers who explain how their bosses cheat them out of wages and workers' compensation, firing those who complain. To highlight objectification, PETA quotes a farm worker as saying he felt he was "disposable" and treated like a "machine," and a contract chicken farmer as saying she was "treated like a dog" by the industry. To further emphasize worker mistreatment, PETA shows pictures of working class people protesting and striking and describes industry as anti-union. In this section, PETA also occasionally uses trigger words for exploitation like *serfs, slaves* and *child labor*. This is contrasted with wholesome "community" values of rural America, or the "heartland," where people simply expect basic, fair treatment from employers and a safe, healthy environment for their families and community. FS's section on the economic issues of factory farming laments the loss of family farms, saying that "small farms help to create close-knit communities and thriving local economies."

Perhaps surprisingly for such a liberal ARO, PETA's sections on the polluting of rural communities and the negligence of government might also appeal to politically conservative values, especially those associated with mistrust of the federal government. Because most AROs propose a consumer solution instead of a government solution, this could be construed as valuing the notion of personal responsibility, consumer choice and free-market capitalism. For example, PETA's page on government negligence shows a photo of the Capitol building in Washington, D.C., specifically emphasizing *federal* government agencies, such as the USDA and EPA and not implicating local governments. Regulation is portrayed as a joke because money has corrupted the process, so consumers, more so than government, must right the wrong by boycotting factory farming.

American Pride

In a few cases, American pride is directly referenced through the use of patriotic symbols, such as the American flag on PETA's bumper sticker bearing the slogan "Proud to be a Vegetarian American." And FARM's Great American Meatout campaign uses red, white and blue colors. One of its posters has Uncle Sam, portrayed by a cow, pointing at the audience, reminiscent of the iconic war recruitment poster, saying, "I want you to stop eating animals." The text emphasizes loyalty by stating that viewers should join the meatout "for your honor, for your family, for your country, and for your planet."

Sometimes the AROs give an indirect nod to American pride by suggesting that the humane policies of the U.S. government lag behind those of other countries. FS uses this strategy of comparing humane laws internationally more frequently than other AROs because it has some of the only campaigns calling for federal legal reform of industry. FS's *Eggribusiness* video explains that European nations have already outlawed battery cages, so "it's time for birds to be protected from abuse in America too." And in FS's *Life behind Bars* video, spokesperson Mary Tyler Moore informs viewers that legal protection for American farmed animals is "grossly inadequate." She states that gestation crates, battery cages and veal crates should be banned in the U.S. as they have been in Europe. The call-to-action is that America has an "ethical obligation" to prevent animal suffering as a "civilized nation." In a similar appeal to advanced civilizations, FARM uses a caveman analogy to imply that meat-eating in the twenty-first century is barbaric.

Personal Values

Health

All AROs except VO prioritize human health as a major benefit of vegetarianism; it is second only to showing compassion for NHAs. Messages tend to be about how a pure vegetarian diet can be healthy in general and often healthier than a standard meat-based diet, especially in preventing major diseases and obesity. They often cite the American Dietetic Association's positive position on vegetarian diets. AROs do not just attempt to say that plant-based diets are *as* healthy as animal-based diets; they often attempt to problematize animal-based diets as *un*healthy. For example, while their health information is mostly positive, COK's and FS's veg guides both say animal products are the "main source of saturated fat and the only source of cholesterol" for most Americans. FS links excess protein intake with a variety of common diseases as well as revealing "links between animal food consumption and many forms of cancer." FS's vegetarian guide also debates the bone-building myth of dairy by saying that "studies suggest a connection between osteoporosis and diets that are rich in animal protein" due to calcium being leached out of the bones. Both COK and FS's guides also list the antibiotic-resistant bacteria strains that are found in animal products, and FS's brochures warn against "harmful pathogens like Salmonella and E. coli" as well as declaring that mad cow disease and avian influenza are "sickening and killing" people.

PETA is the only group that openly appeals to our desire to be sexually active and attractive. On goveg.com, PETA cites a scientific study claiming that meat eating leads to impotence. PETA also takes a more positive approach to sexual enhancement claims by saying that a vegetarian diet helps one to be thinner and more energetic, which is seen as sexier than being overweight and sluggish. This positive association between vegetarianism and sex is endorsed through the organization's annual "sexiest vegetarian" contests. PETA's veg guide has a page on weight loss on which a medical doctor states that vegetarian diets are the "only diets that work for long-term weight loss" and that "meat-eaters have three times the obesity rate of vegetarians and nine times the obesity rate of vegans."

Choice

Through emphasizing choice, AROs appeal to our desire to have plentiful options that can satisfy both the palate and the conscience. COK's materials repeatedly empower people through use of the word *choice,* such as in asking consumers to "choose vegetarian" or "tryveg.com." Choice relates to freedom in terms of not being bound by restrictions on food even though, ironically, veganism is about *not* eating certain things. Yet PETA's goveg.com "Veg101" section declares that vegetarians eat "whatever we want," which is an unusually liberating phrase that implies the choice to eat vegetarian foods is a satisfying preference and not a sacrifice. All AROs illustrate this abundance and variety of food choices with images of bountiful produce, packaged store-bought foods and home-cooked meals.

AROs sometimes remind us that we humans, unlike NHAs, have the luxury to choose whether or not to be conscientious consumers. COK's brochures on eggs and pork both say these animals "don't have a choice—but you do." Similarly, COK has a television spot called "Choices" that asks, "Would you choose to live like this?" as it shows crated animals. In rare cases, it is the NHAs who plead with viewers to choose vegetarian, such as in a few of PETA's collateral materials where the farmed animals say, "please don't eat us."

Pleasurable and Convenient Food

Every ARO highlights the positive aspects of vegan foods, recognizing that taste, convenience, accessibility and variety are very important to food consumers. For example, the *ease* of the diet is often emphasized, especially by PETA, by stating that many accessible options exist now for vegetarians. PETA's starter guide explains, "restaurant options for vegetarian diners keep getting better and better," and "you can now find veggie burgers and other mock meats and soy milk in pretty much every supermarket nationwide, including Wal-Mart." Equally optimistic, COK and FS also declare that it's "easier than ever" to go veg.

To create a positive connotation for vegetarian foods, AROs often accompany messages with cheerful, bright colors such as green, yellow and blue, connoting freshness, and showcasing photos of ripe produce, name-brand convenience products and hearty cooked dishes of common favorites. Frequent use of words like "tasty" and "delicious" imply you won't compromise on taste. The recipe section of COK's starter guide, labeled "Recipes for Vegetarian Delights," assures readers that "eating vegetarian foods doesn't mean giving up the tastes you love." And FARM also uses positive marketing when enticing readers to sign up for its Meatless Mondays campaign: "Have fun. Remember, going veg isn't about restricting your diet—it's about discovering new possibilities and experiencing fresh, exciting flavors."

Belonging (Especially to the Right Crowd) or Desire for Popularity

All AROs emphasize the growing popularity of vegetarianism, presumably so it does not seem like a fringe lifestyle or odd dietary choice. People do not want to be alienated, so, by emphasizing popularity, AROs provide assurance that vegetarianism is socially validated. To combat hippy stereotypes, FS's starter guide assures readers that there are a wide variety of people who eat vegetarian, saying, "from former cattle ranchers to Hollywood celebrities, more and more people from every corner of America are recognizing that vegetarianism is good . . . " and "after years on the fringe, meat-, egg-, and dairy-free fare has earned a well-deserved place in the American food culture." One page is dedicated to proving vegetarians are in "good company," as the "best people" have gone vegetarian for ethical reasons, showcasing famous vegetarians throughout history.

PETA often appeals to our desire to be part of the "it crowd." Celebrity pictures and quotations are used to demonstrate that vegetarians are morally progressive, healthy, attractive and popular. PETA's teen booklet features attractive young stars under the headline "everyone's doing it." To further emphasize that beautiful people go vegetarian, PETA hosts annual "sexiest vegetarian" contests and uses naked or scantily clothed bodies in some campaigns.

Integrity, Including Moral Consistency and Pride in One's Morality

I define moral consistency and integrity as reflecting one's values through actions and applying those values uniformly and fairly in all situations. Regarding the major value of compassion, AROs activate the logic of moral consistency as follows: If people already care about the welfare of cats and dogs and do not want to see them harmed, and if farmed animals are equally sentient, then it would make sense that compassionate people would not want to see farmed animals harmed either. To show consensus for farmed animal welfare values, vegetarian guides for FS and PETA both use survey data to prove that most Americans are in favor of legal protection of farmed animals and against intensive confinement. But although a consensus clearly doesn't exist in favor of saving farmed animals from death and consumption, there is a consensus that people should not eat dogs and cats, so that is where AROs often point out moral inconsistencies in American attitudes.

Messages by FS, FARM, PETA and COK use questions as a tool to provoke viewers to rationally justify why they *eat* certain species and *befriend* others, implying it is a morally random decision who gets killed. A COK t-shirt shows a photo of a dog seated on a dinner plate with a knife and fork on either side. The headline asks, "Why not? You eat other animals, don't you? Go vegetarian." Similarly, a FARM vegetarian postcard shows a picture of a cat and a piglet nose to nose with the question "Which do you pet? Which do you eat? Why?" FS collateral materials show a cartoon of a happy dog and cat and an anxious cow and pig with the question "If you love animals called pets, why do you eat animals called dinner?" And to help create empathy for sea animals, a PETA brochure says we humans wouldn't "stab our cat or dog through the mouth" (as a fishing analogy), and "none of us would drop a live cat or dog into boiling water. Why should it be any different for lobsters?"

VO also appeals to moral consistency, as their booklets openly talk about the need for people to widen their "circle of compassion" to include farmed animals. The *Even if You Like Meat* booklet states that most people are "appalled" by farm animal cruelty, not because they believe in "animal rights," but because they "believe animals feel pain and that morally decent human beings should try to prevent pain whenever possible." In this way, the appeal is not asking for a *change* in values, since it assumes people are generally supportive of animal welfare, but rather it asks for an equal application of this welfare value.

By undermining the welfare claims made by some free-range farms, AROs attempt to show that veganism is a true reflection of one's compassion, while supporting so-called "humane" farming is not. For example,

FS's position paper on "Humane Meats" says that people who are "sincere" in their concern for animals will stop eating them, insinuating that animal lovers are hypocritical if they still eat meat, even "free-range."

FRAMING IMPLICATIONS AND RECOMMENDATIONS

By appealing to a wide variety of culturally relevant values, AROs largely follow the strategic framing advice for social movements for creating an appealing identity—in this case one that is broad enough to dispel myths that vegetarians represent only a feminine subculture of liberal hippies and animal rights activists. And while the diet they promote is vegan, they usually call it by the more open and familiar term *vegetarian,* per Maurer's findings. By appealing to people's *values* instead of just reason, AROs follow Lakoff's advice, but they don't necessarily provide a collective moral vision, such as of a vegan world. Considering that over 95% of the population eats animal products, promoting a vegan America may be too utopian at this point. So AROs settle for displaying optimism about a movement toward vegetarianism, which provides a vision for a better world as each vegetarian helps mitigate global problems, such as animal suffering, disease and obesity, pollution and climate change, resource depletion and human hunger. In this effort, most AROs promote the three standard vegetarian tenets of animal ethics, human health and environmentalism, in that order. But most do not follow the Humane Research Council's pragmatic advice to promote meat-reduction ("Advocating Meat"), with the exception of FARM's Meatout Mondays campaign and VO's *Even if You Like Meat* booklet. And even in these latter approaches, a transition to veganism is still touted as the best option. In keeping with an animal rights mission, AROs do not suggest that Americans switch to so-called "humane" animal products, preferring that people eat plants, not better-treated animals.

AROs mainly align their values with those of the audience by using frame building, amplification and extension processes (Snow et al.). Companion animal welfare is used as a tool for frame *bridging;* AROs often use analogies comparing the sentience of farmed and companion animals in an attempt to use logic and a plea for moral consistency to get the public to transfer to land-based farmed animals their respect for the subject status and individuality of dogs and cats. In selecting species, most AROs emphasize mammals over birds and fish, even though the latter are killed in greater numbers, as humans can presumably identify more with fellow mammals who more closely resemble themselves and their companion animals. PETA was the exception, as they attempted to build that longer bridge to connect people to aquatic animals.

AROs implement frame *amplification* by deepening the notion of what it means to be compassionate and respectful toward NHAs, so that these values applied to supporting freedom and life values (veganism) instead of just reducing suffering (humane farming). AROs also amplify appeals to American populism and corporate skepticism; they apply this appeal to animal agribusiness, in particular factory farming, stating that it is cruel to NHAs, destructive to the environment, unfair to human workers and misleading to consumers. Related to the populist ideal, frame *extension* occurs where AROs equate vegetarianism with human rights and environmental causes.

My recommendations for increasing the resonance and logical consistency of ARO frames include an emphasis on justice, freedom and life, and a reframing of naturalness and honesty. To begin, I contend that the AROs implicitly appeal to the American values of rights and justice when they use appeals to compassion and respect for animal subjectivity as a call-to-action for veganism (animal rights) instead of for a switch to so-called "humane" animal products (animal welfare). More direct comparisons between human and nonhuman animals and between human rights and animal rights would further bolster this abolitionist stance in favor of justice, escalating the frame alignment process from amplification to transformation (ideological change).

While AROs generally appeal to *naturalness* by framing factory farming and slaughterhouse practices as "unnatural" in comparison to traditional animal farming, AROs could more logically promote veganism by extending the naturalness frame to communicate that *all* farming of other animals for food is itself unnatural when viewed in relation to how predation operates in nature through hunting and scavenging. This connects naturalness to *freedom*, indicating that all animals, human and nonhuman, want to live free of captivity.

Overall, AROs appeal to the best of our humanity to show how veganism is a natural fit with many cherished values. Therefore, I contend the emphasis on *moral integrity* is one of the most crucial appeals in creating a vegan upsurge, as attitude and behavior changes likely hinge on creating cognitive dissonance over meat's fit with one's value system (Joy 141). The appeal to *honesty* could be tied more directly to moral integrity if it focused less on expecting truth from the meat industry and more on humans being truthful with themselves. An honesty frame could state that one should willingly, openly and frequently confront the agricultural practices and consequences behind one's food choices to ensure they are in accordance with one's own values and to maintain moral integrity and model it for one's children.

CONCLUSION

According to these animal rights organizations, veganism says a lot about someone. It says they likely prioritize altruistic values such as: compassion, respect for sentient beings, life, freedom, environmentalism and the desire to make a difference and help humanity. It says they likely believe in ideals such as: honesty, naturalness, patriotism and populist notions of fairness toward the little guy. It also says they identify with common-sense personal values such as: health, choice, belonging and social appeal, desire for pleasurable and convenient food and pride in one's moral integrity.

Whereas meat-industry executives or wing eaters at a sports bar might disagree with this positive and admittedly flattering characterization of vegans (as meat eaters might see many of these values as applying to themselves and not conflicting with meat eating), the animal rights movement seeks a broad identity for vegans in order to make veganism a mainstream dietary choice. Perhaps that is why the appeals seem bipartisan, non-gender-specific and nondenominational. AROs attempt to show how a variety of values with which many Americans already identify are a natural fit for choosing plant-based foods and are out of sync with their current habit of consuming animal products. But if plant-based foods represent altruism, health and sociological and environmental responsibility, then do meat, eggs and dairy represent the opposite? Is America's food identity represented by selfishness, irresponsibility, unsustainability, violence, injustice and apathy toward the lives and suffering of humans and other animals? If so, this hypocrisy is concealed by the dominant commercial discourse of pleasurable consumption because most Americans would not identify with these unflattering traits. Meat eaters might argue that they simply frame or prioritize their values differently when it comes to food.

But you do not see AROs constructing a heartless meat eater as much as they choose to construct a thoughtful vegan. Blame for problems is mainly targeted at industry, while AROs choose positive and optimistic appeals toward consumers as the solution. The underlying subtext suggests, "Now that you know about the problems with animal agribusiness, an ethical and rational person such as yourself will surely make the right choice and go vegan." But once consumers *do* know about the suffering, injustice, pollution, etc., if they still choose to continue to support it, how is the animal rights movement supposed to address them? What if consumers do not mind being hypocritical or do not agree with extending animal welfare or rights values to "food" animals? Will meat eaters be increasingly shamed as part of the problem?

Regardless of whether AROs give the public the continued benefit of the doubt regarding blame, meat eaters ultimately hold the solution. So, appeals will continue to be made for them to literally and metaphorically step up to the veggie plate, hopefully not just as individualistic food consumers but also as altruistic citizens who are politically and socially engaged in improving society and living their values—one bite at a time.

REFERENCES

"About FARM." About FARM. N.d. Web. 14 Nov. 2007.
"About COK." *Welcome to Compassion over Killing.* N.d. Web. 14 Nov. 2007.
"About FS." *Farm Sanctuary: About Us.* 2008. Web. 11 Apr. 2008.

"About PETA." *PETA's History: Compassion in Action.* N.d. Web. 14 Nov. 2007.

"About VO." *About Vegan Outreach.* N.d. Web. 14 Nov. 2007.

Advocating Meat Reduction and Vegetarianism to Adults in the United States. Humane Research Council report, 2007. http://www.humanespot.org/node/1956 Web. 8 Apr. 2008.

Benford, Robert D., and David A. Snow. "Framing Processes and Social Movements: An Overview and Assessment." *Annual Review of Sociology* 26 (2000): 611–639. Print.

"Farm Death Toll." *World Farm Animals Day.* 17 Oct. 2007. Web. 29 May 2009.

"FS Financial." *Farm Sanctuary, Who We Are,* and *2006 Form 990.* Guidestar. org. 2006. Web. 14 Nov. 2007.

Gamson, William A. *Talking Politics.* New York: Cambridge UP, 1992. Print.

Gitlin, Todd. *The Whole World Is Watching: Mass Media in the Making and Unmaking of the New Left.* Berkeley, CA: U of California P, 2003. Print.

Hall, Stuart. "Introduction." *Paper Voices: The Popular Press and Social Change, 1935–1965.* Ed. A. C. H. Smith. London: Chatto and Windus, 1975. 11–24. Print.

———. *Representation: Cultural Representation and Signifying Practices.* London: Sage Publications, 1997. Print.

Johnston, Hank, and John A. Noakes. *Frames of Protest: Social Movements and the Framing Perspective.* Lanham, MD: Rowman and Littlefield, 2005. Print.

Joy, Melanie. *Why We Love Dogs, Eat Pigs, and Wear Cows: An Introduction to Carnism.* San Francisco, CA: Conari Press, 2010. Print.

Lakoff, George. *Don't Think of an Elephant!: Know Your Values and Frame the Debate.* River Junction, VT: Chelsea Greene Publishing, 2004. Print.

Maurer, Donna. *Vegetarianism: Movement or Moment?* Philadelphia, PA: Temple UP, 2002. Print.

Peak Meat: U.S. Meat Consumption Falling. Earth Policy Institute report. March 7, 2012. http://www.earth-policy.org/data_highlights/2012/highlights25 Web. 2 Oct. 2012.

Pollan, Michael. *The Omnivore's Dilemma: A Natural History of Four Meals.* New York: Penguin Press, 2006. Print.

Singer, Peter, and James Mason. *The Ethics of What We Eat: Why Our Food Choices Matter.* 1st ed. Emmaus, PA: Rodale, 2006. Print.

Snow, David A., and Robert D. Benford. "Ideology, Frame Resonance and Participant Mobilization." *International Social Movement Research* 1 (1988): 197–218. Print.

Snow, David A., E. Burke Rochford, Steven K. Worden, and Robert D. Benford. "Frame Alignment Processes, Micromobilization, and Movement Participation." *American Sociological Review* 51.4 (1986): 464–481. Print.

Stewart, Charles J., Craig A. Smith, and Robert E. Denton, Jr. "The Persuasive Function of Social Movements." *Persuasion and Social Movements.* 4th ed. Longrove, IL: Waveland Press, 2001. 51–82. Print.

Tarrow, Sidney. *Power in Movement: Social Movements and Contentious Politics.* 2nd ed. New York: Cambridge UP, 1998. Print.

Zogby, Joseph. *Nationwide Views on the Treatment of Farm Animals.* Animal Welfare Trust, 2003. Web. 28 May 2010.

7 The "Golden" Bond
Exploring Human-Canine Relationships with a Retriever

Nick Trujillo

"Ebbet is very much looking forward to the trip," Denise Shultz told me during lunch in Occidental, California, a few days before Ebbie and I began our journey.

"He considers himself an ambassador on a mission of peace," added Winterhawk, Denise's business partner.

Ebbet, or "Ebbie" as I usually call him, is my golden retriever, and Denise and Winterhawk are dog psychics, or animal communicators, as those in the industry prefer to be called. Ebbie and I had taken a day trip from our home in Sacramento to Occidental, a bohemian village near Bodega Bay, to visit with them before we embarked on a five-week journey up the California coast. Ebbie and I would be on a quest to find dogs and their owners as well as people who work in dog-related businesses, and I had emailed Denise and Winterhawk to set up an interview for the following month. They agreed to the interview, but only if they could meet Ebbie before the trip to find out what his expectations were.

I am skeptical of the animal communicator industry, and yet I was strangely pleased when they said Ebbet was looking forward to our tour of the Golden Coast. The idea that this would be a "mission of peace" was also comforting, even inspiring. Like John Steinbeck and his poodle Charley forty years earlier, the "Ambassador" and I were hitting the road to learn about our fellow mammals, though Steinbeck spent three months traveling across the U.S. and Ebbie and I had only five weeks to explore the California coast. Steinbeck went in search of American culture, but I wanted to learn about *dog culture*, the shared values and beliefs of 'dog people' and their canine companions. Unlike Charley, who merely went along for the ride, Ebbie was a full participant-observer, sniffing and licking other dogs and their human companions.

We ultimately met hundreds of dogs and their owners. Before the trip, I lined up interviews with about fifty people who worked with dogs, including dog trainers and dog walkers, owners of dog bakeries and dog boutiques, K-9 officers and Customs agents, volunteers who handle therapy dogs and disability dogs, the writer of a dog book and the editor of a dog magazine and even the owner of a dog-poop pickup service. I also interviewed and observed over one hundred dog owners at dog parks and beaches along the way. I asked a variety of questions, including: Why did you get a dog? Why this particular dog? What do you do with your dog? How do you and your dog communicate with each other? Do you ever talk to your dog? If so, what do you say to him or her? How would you characterize the relationship you have with your dog? These and other questions, adapted depending on the position of the person interviewed, elicited various stories and accounts about the nature of humans' relationships with their canine companions.

This chapter offers some preliminary insights into the nature of human-canine relationships based on this five-week study of dog culture. I discuss the variety of human-canine relationships that I found, the relevance of various theories of interpersonal relationships to explain them and differences between men and women with respect to human-canine relationships. I end with personal reflections on the nature of my relationship with Ebbie and with a postscript about life without him.

THE VARIETY OF RELATIONSHIPS
BETWEEN HUMANS AND CANINES

Humans have a wide variety of relationships with others, including coworkers, acquaintances, friends, family members and lovers. During my five-week exploration of the Golden Coast, I found that humans and canines have similarly diverse relationships. The best-friend cliché does capture one type of relationship for many people. Throughout the trip we met several people who described their dogs as "friends," "pals" and "buddies": "Buddy is my buddy," a retired Newport Beach resident named John said, describing his Sheltie. "You can have a bad day and get angry at the dog when you're really angry about something else, and a minute later the dog has forgiven you for yelling at him."

I also met several others for whom their dog was a "coworker" or "working partner." Indeed, all of the police officers, customs agents, rangers and other officials who worked with canines were partnered with a particular dog. "In the old days they had multiple handlers," a ranger at Hearst Castle told me. "You might have one dog and four officers. Along the way they decided that wasn't the best way to do it. The one-on-one bond between the dog and the officer is better." In most cases, this "bond" between an officer and his or her dog is powerful. For example, at the War Dog Memorial in Riverside, I met a Vietnam veteran who handled a scout dog named Ken for eighteen months during the war. At one point he asked me if I had ever eaten dog food. I admitted that I ate some dog biscuits once when I was a little kid.

"Well, there were times on patrol that we'd run out of rations," he said. "So you're eating your dog's food. Then other times you run out of dog food and your dog is eating ham and lima beans. Let me tell you, when you're under a poncho in the rain out on patrol, and you've both eaten ham and lima beans, the smell can get pretty ugly," he said and laughed. "That gives you an idea of the bond we had with our dogs over there." Sadly, the U.S. government euthanized the military dogs at the end of the Vietnam War, a decision that military dog handlers will never forget. "That's why this is so important," he said, pointing to the War Dog Memorial. "It gets the word out that the United States government killed our dogs." In November 2000, Congress passed Resolution H.R. 5314, an act that promotes the adoption of military dogs at the end of their service and requires an explanation whenever a military dog is euthanized rather than retained for adoption. This act does not prohibit the killing of military dogs in combat situations, however, because they still are considered "tools of war." Organizations such as the Military Working Dog Foundation and others do everything they can to assure that retired military and other working dogs are adopted rather than euthanized.

Ebbie and I traveled to Oceanside to visit a different kind of service dog at Canine Companions for Independence. Founded in 1975, CCI is a nonprofit organization that provides trained assistance dogs to people who qualify. Their dogs open doors, turn lights on and off and perform other tasks for people with disabilities and alert people with hearing loss to telephones, alarm clocks and other sounds.

Ebbie and I met Rosemary, an eleven-year-old girl in a wheelchair, and her canine companion Eureka, a black Lab that never left her side.

"Eureka is the most greatest dog I've ever had," Rosemary said and smiled widely. "He is very special to me. He's very reliable and very intelligent."

"What does he do?" I asked.

"He can retrieve pens and pencils," she said. "When I do my homework, they roll off the table. He can turn lights on and off, push the handicapped button to open doors and open and close drawers. He protects me at night, and he shows me unconditional love."

Rosemary reached down to pet Eureka and Eureka leaned in, anticipating her touch. I could see they shared a powerful bond, although I couldn't imagine how such a connection would feel. When I was her age, I thought of my dog Casey as the brother that I never had, but he stayed home when I went to school and to play with my friends. Eureka goes *everywhere* with Rosemary, and their relationship is a fitting testament to CCI.

Many of the everyday dog owners we met during the trip described their dogs not simply as "friends," but also as members of their families. "We take Toy and Shadow everyplace," said an older couple about their Shih Tzus. "To the grocery store, on trips, everyplace. They're part of the family." "People who don't treat dogs like family members shouldn't get dogs," the owner of a "free range" doggie daycare center in L.A. called The Loved Dog told me. "They lack the ability to empathize, to connect."

Although there may be some people who fit the stereotype of the "cat lady" or "dog guy" who seem to spend time exclusively with their animals, I found evidence that dogs *help* rather than hinder the development of relationships among humans. Ebbie and I spent one afternoon at a little park on "The Strand" near Hermosa Beach Pier, as hard-bodied men and women strutted alongside middle-aged men with beer guts and middle-aged women with sagging breasts, and vice versa. On The Strand, Ebbet definitely was an ambassador, if not of peace then of communication. Numerous people came up to pet him and to say hello, confirming the fact that dogs are effective facilitators of human interaction. People who might not have made eye contact with me if I were by myself stopped to chat. They saw Ebbie first and almost always smiled, and then greeted me warmly.

In her remarkable book *Pack of Two*, Carolyn Knapp reports research finding that people who have dogs are perceived to be friendlier, happier and less threatening than those who don't have dogs. Our experiences on The Strand certainly supported those results. I could have been a serial killer, but with a pretty dog at my side I looked trustworthy. Knapp also argued that dogs widen rather than narrow one's social circle, especially for people who take their dogs on walks and to parks. By facilitating interaction among people, dogs increase our chances of developing friendships and romantic relationships with each other.

A twenty-something man named Jamie hoped his eight-week-old puppy Maya would lead to the latter. The two of them attracted quite a crowd at the park, and he told me that he had wanted to get a dog for a long time. I asked what finally motivated him to get one. "I realized that she's a great babe magnet," he said and smiled. "I get a lot of compliments from guys too, but they don't usually want to hold her like the girls. I guess it's some kind of unwritten rule." Sure enough, a few minutes later, two young women approached.

"Can we pet your puppy?" one of them asked and bent down to touch her, exposing even more of her loosely-bikinied cleavage to Jamie.

"Sure, ladies," he said and winked at me.

I laughed, thinking how manipulative it was to use a dog to attract romantic partners. Then I remembered that my wife once admitted purchasing her first Airedale Terrier in part to meet men. I thought that sounded pretty desperate at the time, but Jamie's disclosure suggested that more people might use dogs to get dates than I ever imagined. Perhaps that explains why, as Marjorie Garber wrote, the expression "walking the dog" is a slang term for having sex.

After several minutes of puppy love, the women moved on and Ebbie and I went to The Strand on the Manhattan Beach side. I sat on an adobe fence in front of a condo and Ebbie lay on the concrete, waiting for dogs and dog people.

A woman in her late sixties named Margaret and her yellow Lab-mix named Buster stopped by for a chat. Margaret was a widow and admitted that she had become much closer to her dog since her husband died almost two years earlier. "Buster was really Hal's dog," she said. "But we've bonded in the last year and now we go everywhere together."

Her statement reinforced the idea that dogs help people relate to each other, though it seemed to represent the other side of the relational continuum. The young man in Hermosa bought a puppy to meet women and start a relationship, whereas the older woman in Manhattan remembered her deceased husband through their old dog. I watched Margaret scratch Buster and think of Hal, and I realized how much dogs help humans connect with each other throughout our relational life cycles. This was a lesson I would learn firsthand a few years after the trip.

CONCEPTS FOR UNDERSTANDING
HUMAN-CANINE RELATIONSHIPS

Some of the concepts and theories that scholars in communication use to describe and explain interpersonal relationships may also be appropriate for characterizing the human-canine relationship, whereas others may not be appropriate. Obviously, any theory that relies on *verbal* communication will not be especially useful, even for people who have taught their dogs to "speak." For example, although self-disclosure is a key concept for understanding how people develop relationships with other people, it may not be especially useful for understanding human-canine relations.

This is not to say that people do not speak to their dogs. Many people I met on the trip do so regularly. One woman I met said she says "good morning" and "good night" to her dogs every day, and talks with them throughout the day. I also speak to my dogs regularly, and they understand at least a dozen or so words, including, sit, stay, roll over, lie down, leave, heel and of course doggie din-din. I also talk with my dogs—simple things like "How's it going Boys?"

Other concepts, however, may help us understand human-canine relationships. For example, the needs theories of William Schutz and Abraham Maslow provide insights into the motives behind and nature of our relationships with dogs. Schutz argues that three basic interpersonal needs are satisfied through relationships with others, including inclusion (the need to be involved with others), affection (the need to have positive feelings from and toward others) and control (the need to influence and be influenced by others). Certainly all three needs help explain some of the motives and nature of our relationships with dogs. The need for inclusion manifests itself in a dog's desire to be in a pack, and that pack can include not only other dogs but also humans; as Jeffrey Moussaieff Masson writes in *Dogs Never Lie about Love*, "no other species has ever indicated that it regularly prefers the company of a human to that of members of its own species, with the single exception of the dog" (15). The dog lovers that I met on the trip said they also feel a need to be with their dogs and miss them when they are not with them.

The need for control manifests itself in a dog's desire to know his or her place in the social hierarchy of the pack, and this desire is communicated among dogs with a fairly sophisticated set of signals that indicate dominance and submissiveness. People also have a need to control their dogs through formal or informal obedience training, and I met too many people on the trip that needed to exhibit more basic control over their dogs than was sufficient to prevent them from jumping on, sniffing and doing other inappropriate acts to other humans and canines.

Finally, the need for affection manifests itself in a dog's so-called "unconditional love" of humans, though critics may argue that a dog's love is based on many conditions (e.g., to get treats, to get us to play with them, etc.). Some people also expressed a need to care for and love—truly love—their dogs. With respect to Maslow, dogs probably have more of a physiological need for humans than vice versa, though if we didn't feed them, they could certainly fend for themselves as feral dogs. Dogs and humans satisfy security needs in a mutual way: People protect skittish dogs and many people buy or adopt dogs to give them more security. Dogs' and humans' social needs were discussed above through the lens of Schutz's affection and inclusion needs. Ego needs may be fulfilled through inclusion, too, though they may also be satisfied for people who cultivate their identity as a "dog person" or develop an identification with a particular breed, as illustrated by many of the enthusiasts that I met at a dog show in Pomona. I doubt that a dog cultivates a similar identity as a "person canine" or as a dog who enjoys being with a person of a particular race or gender. I did meet a man who insisted that his dog "did not like black people," but customs agents and therapy-dog handlers said that dogs are effective precisely because they are not biased and are not bothered by anyone's mental or physical attributes. I'm confident that the man's dog sensed *his owner's* discomfort around African-Americans and responded protectively. Anyone who argues that their dog is racist clearly doesn't understand dogs or their own racism.

Finally, there is self-actualization, that rarely achieved state of sublime consciousness. Do dogs experience such a state through their relationships with humans? I suppose only dogs (and perhaps animal communicators) know for sure. I did, however, meet some dog lovers who suggested that they experience at least a limited sense of self-actualization through their dogs. For example, I talked with a therapy-dog handler who took her beagle Ruby to visit a four-year-old girl in a coma. The parents had told her that their daughter loved dogs. "We went into her room, and I put Ruby on the little girl's bed," she said. "A nurse took the little girl's hands and had her pet Ruby. And just then the little girl's eyes opened."

Goosebumps erupted on my arms.

The nurse took a Polaroid picture of the little girl with her eyes open before they closed again. Ann saw the mother and father about an hour later. "They had tears in their eyes," she said, blinking back her own. "They were holding the photograph of their daughter with her eyes open." Now I was blinking back tears, moved by the power of a dog and a little girl's love of dogs.

A variety of other interpersonal concepts may offer additional insights into human-canine relationships. Social exchange theory might suggest that we will keep a dog only when the benefits outweigh the costs. Unfortunately, far too many people abandon and/or abuse their dogs when they grow up and no longer provide the cuddly benefits of puppyhood to people or their children, something I saw firsthand when we visited the San Diego Society for the Prevention of Cruelty to Animals. Perhaps concepts such as abuse, anger, aggression and others from the "dark side" of interpersonal communication, studied by Brian Spitzberg, William Cupach and others, may be useful in understanding the dark side of human-canine relationships.

The concept of attraction—task, social and physical—may also be relevant in some ways. Several K-9 officers said they cherished working with their dogs; most of the people we met enjoyed spending leisure time with their dogs, and, not surprisingly, most owners thought their dogs were "pretty" or "beautiful." But are dogs attracted to us? Certainly the working dogs seemed to enjoy working because, as customs agents at the border explained, such work is play for them. Everyday dog owners insisted that their dogs like going to beaches and parks with them as well. I don't know if dogs are attracted to us physically, but they certainly make us feel attractive when we come home and they run to greet us and lick our faces.

Uncertainty reduction is another concept that might have some potential to explain human-canine relationships. Charles Berger and C. J. Calabrese argue that when people meet, they seek to reduce the uncertainty they have about each other. They do this by observing and interacting with each other. Obviously, humans and dogs need to reduce uncertainty about each other. Those of us who take our dogs to training classes are taught to read their nonverbal communication to get to know how they will respond, while they learn to read our nonverbal and verbal commands. And, of course, dogs use their acute sense of smell to greatly reduce uncertainty in their lives.

Another interpersonal concept is attitude similarity. I don't know of anyone who has studied the attitudes of dogs, but conventional wisdom suggests that some people are similar to their dogs. When Ebbie and I stumbled across a walk for cystic fibrosis in the sleepy coastal town of Carlsbad, we joined the crowd at the finishing line in Magee Park, and I was struck by how much some people and their dogs look alike. A woman with long red hair parted down the middle walked with her Irish setter. A flamboyantly dressed male strutted with his show-cut poodle. And a hulking man with a shaved head lumbered by with two huge Rottweilers, each weighing well over 150 pounds. They certainly looked the same, though I don't know if they shared the same attitudes.

We'll never know if dogs and people share similar attitudes, but most dog lovers have played the human game of "what kind of dog are you?"—a game that definitely taps into attitudinal dispositions. Friends who knew me and my wife agree that she was a terrier and that I am a retriever.

Finally, there is the concept of relational development. Mark Knapp and Anita Vangelisti argue that humans develop relationships in stages characterized by increasing levels of self-disclosure and intimacy until they achieve "bonding," the final state in relational development during which the people in the relationship commit to each other. Although humans and canines do not self-disclose or commit to each other in the same way, relationships between dogs and people definitely progress through stages and can result in a unique and powerful form of bonding.

GENDER DIFFERENCES IN HUMAN-CANINE RELATIONSHIPS

The cliché "man's best friend" is not only sexist; it is also limited. Many women develop close relationships with dogs. Ebbie and I met numerous female owners of dogs, and most of the managers and employees of dog businesses, especially small niche businesses such as dog boutiques and dog bakeries, were women. The female co-owner of Fideaux, a dog boutique in Carmel, agreed that most people in her profession are in fact women, though she said that the owner of George, a dog shop in San Francisco, is male.

"Of course, he's gay," she said and smiled.

I wondered what attracts females to this work and deters males from it. I suppose it is more macho to sell computers and used cars rather than doggie biscuits and cashmere sweaters for Chihuahuas.

But do women treat dogs differently than men and develop different relationships with their dogs than do men?

Women seem to dress up their dogs more than men do. At a fund-raising event for a humane society in Rancho Santa Margarita, Ebbie and I met several women and their dressed-up dogs, including a dog groomer who had dyed her white standard poodle red, white and blue, though the red was pink and the blue was baby blue. At the same event, we also met members of a dancing dog troupe called "Muttley Crew," including Mary and her border collie Tory. Mary wore a wizard cap and cape and Tory wore a feather vest and a polka-dotted skirt. One of the dog stores that Ebbie and I visited was Fifi and Romeo, a boutique on Beverly Boulevard where affluent women buy clothes, bags and other canine accessories. "Putting clothes on your dog is not a novel thing," the co-owner said. "But we approached it from a fashion point of view. Does a dog need a cashmere sweater with hand embroidery? Of course not. But being a girl, I wouldn't wear an acrylic sweater, so why put one on my dog?" She added that her customers usually have small dogs. "They're small, almost like accessories," she said. "So we designed this as a handbag that can sneak your little dog into restaurants and other places."

Apart from these cosmetic distinctions, there may be other differences between male and female dog owners. A feminist critic I interviewed summarized the reasons that she had a terrier. "They empower you," she said.

In this society women do not have as much power as men. It is more uncomfortable for women to walk around a city simply because we are women. So having a dog makes you feel safe and more in control. They also give you something to nurture. I don't have children, and my dog allows me to enjoy the nurturing that I would have experienced with kids.

On the other hand, when I asked men why they had dogs, the "buddy" theme came up consistently. And, as one man put it, "He's someone I can spend time with who won't talk back or ask me to talk about my feelings." Interestingly, however, men also disclosed that they feel affection for their dogs. "I loved that dog more than some of my relatives," one man said, talking about his retriever that had died recently.

Ironically, then, dogs may help females feel more powerful (i.e., more "masculine") and help males feel more emotional (i.e., more "feminine"). In this way, dogs not only help people relate to each other, but may also help men to vicariously experience what it is like to be a woman and women to vicariously experience what it is like to be a man.

MY RELATIONSHIP WITH EBBIE

Many of the concepts discussed earlier apply to my relationship with Ebbie, named after Ebbet's Field, the former home of the Brooklyn Dodgers. I named my other Golden Wrigley, after the home of the Chicago Cubs, and friends joke that my next retriever will be named Pac Bell or Qualcom.

Clearly, the concept of inclusion characterizes our relationship. Whether it is a need or an "instinct," Ebbie is very connected to his pack, our family. I don't know if Ebbie thinks of our pack as his *family*, but Leah and I certainly think of Ebbie and Hawkeye as members of our family. They live with us in our home, not outside in the backyard, and "Hawkie" even sleeps in bed with us. We give them presents for their birthdays and Christmas. We know they are not human children, but they are dependents and we love them as deeply as—and in some cases even more than—our relatives. In fact, because Leah and I did not have children, "the boys" are the reason that we see ourselves as a "family" rather than merely a "couple."

Control is another concept that characterizes our relationship, and that became clear when Ebbie and I visited Huntington Beach Dog Beach, an off-leash beach in Orange County. We met many people and dogs, including Carol, her Golden Retriever Max and her Husky-mix Freedom. Max is a Frisbee dog, but Carol's kids forgot to bring his disc. So he and Freedom played with Ebbie in the water, bobbing for tennis balls.

I looked back and forth between Carol and Ebbie, maintaining rapport with Carol but making sure that Ebbie was fine. He was still in the surf with Max and Freedom, along with a lean yellow Lab named Budweiser who had joined them. But when I turned around again, Max and Freedom were there, but Ebbie and Budweiser had vanished. I panicked, thinking that they had been pulled into the sea by a riptide, until I looked to the north and saw them sprinting together up the beach. "Ebbet!" I yelled out. "Halt!" But he just kept running with his new Bud.

I stopped talking with Carol and started running after Ebbie. It was a Leah-would-not-be-happy moment. For Leah, it is inappropriate to lose control of your dog and quite unacceptable to have no idea where your dog is going. She would have been mortified to see that Ebbie was so entirely out of my control.

I wasn't mortified, but I was angry and embarrassed that my dog was ignoring my directives. I wanted to scold him, but, of course, I couldn't catch him. I finally stopped running and watched Ebbie and Budweiser move together. They were totally free of human control, a beautiful sight to behold. They were not racing, just running in concert and banging their bodies into each other in lusty play. I actually was jealous, and at that moment I wished I were a dog so I could run with them.

Ebbie eventually tired out and started to fast waddle, though Budweiser could have continued their steeplechase pace indefinitely. Ebbie stopped at the shoreline and waded into the surf, lapping up several tongues-worth of seawater. "Ebbet, no!" I yelled out, not wanting him to get sick from drinking salt water. "Ebbet!" I shouted again, using his formal name as my mom used mine when she was upset with me in my youth. Ebbie heard me or saw me or smelled me, finally acknowledging my commands. He walked toward me with his head down, knowing I was not happy with him. I wanted to punish him, but I know you shouldn't do that when a dog is coming to you. Budweiser joined us on the long walk back to the crowded area where they started their escape. I found his owner, who said that "Bud-dog" does that every time she takes him to Dog Beach. I kept Ebbie on a very close heel with voice commands, treats and evil glares.

Despite Ebbie's breakaway, we had a great time at Huntington Beach. And because of that breakaway, I recognized, reluctantly, that authority and obedience define a bigger part of my relationship with Ebbet than I like to admit. I like to think of Leah as the control freak in the family because she would rarely let Hawkeye off leash or let him play with other dogs, whereas I allow Ebbie to be off leash and interact with his fellow canines. However, as Ebbet's "owner" and "master," I demand that he comply with my orders, and when he doesn't I reprimand him or bribe him with treats until he does. He eats when I want him to eat, and when he is on leash I demand that he "heel" at my pace. I call him a "good dog" when he obeys me, and a "bad dog" when he doesn't.

Yet I consider myself to be defiant to authority, a trait developed in my adolescence whenever my father ordered me to do something, which was quite often. My dad has threatened to write a book titled *Take Out the*

Trash, an amusingly painful reference to the fights that we reenacted almost every day when he ordered me to take out the trash.

"Take out the trash," he would demand in his autocratic tone.
"Say please," I would counter, knowing that he rarely said please to anyone.
"I don't have to say please!" he would respond, raising his voice.
"If you don't say please, then I don't have to take out the trash!"
"Then you don't have to stay in my house!"
"Then I'll leave!"
"Then go ahead!"
"Okay, I will!"

I would leave the house, slam the door and drive around in the 1960 Rambler station wagon that I inherited on my sixteenth birthday. After cooling down, I would come home and my dad and I would apologize to each other. Then we would repeat the same performance the next day for the sequels *Clean Your Room*, *Give Me the Salt*, *Turn the Channel* and others.

Ironically, my father is also defiant to authority. He hated his brief stint in the military because officers ordered him around, and he was fired by Wayne Newton when the Las Vegas entertainer demanded that my dad laugh at his jokes between songs and my dad responded, "The only thing that's funny about you is your singing." Although my dad resisted the demands of others, he readily made them of me, and when I did not automatically comply, he became defensive and angry, which prompted more defiance from me. And the dysfunctional pattern continued to cycle, and I am resistant to authority to this day.

As I replayed the image of Ebbie and Budweiser frolicking so beautifully without any constraints, I recognized the inconsistency in my tendency to resist authority and my willingness to control Ebbet. I decided to give the Ambassador more freedom during the early part of our mission, allowing him to dictate the pace of our walks, to eat seaweed at will and to explore some of the dog friendly companies with only canine escorts. As a result of my laissez-faire leadership, Ebbie injured his paw on a long walk, threw up after eating too much seaweed and got lost in the dog-friendly production houses of film director Ridley Scott. And I decided that control is good.

"Most dogs really don't want to be the alpha dog," a dog trainer and dog walker in L.A. told me. "They want you to be the leader. They're happiest when the pack is clearly defined and they have boundaries. They're like two-year-old children. But they are *dogs* and they want you to draw a line."

Finally, I have genuine affection and love for Ebbie, and I believe that he has affection and love for me. I have to admit that I wasn't sure if I would ever love a dog quite the same way after my beloved Wrigley died in the mid-1990s. I grew up with many dogs, including Blue, a Kerry Blue Terrier, Nuisance, a Terrier-mix, and Casey I, Casey II and Candy, all Keeshonds.

But Wrigley was my first dog as an adult. I bought him on an Aurora, Illinois, farm about thirty miles from Chicago the day after the Chicago Cubs lost to the San Diego Padres in the 1984 National League Championship. He was the smallest male of the eight-week-old litter and was cowering under a picnic table, and it was clear that his siblings had been picking on him for seven weeks and six days.

I felt sorry for the little guy, and brought him home.

By the time Leah and I married and moved to Dallas one year later to take jobs at Southern Methodist University, Wrigley had become a truly amazing dog. I was a pitcher in high school and college, and I could instruct Wrigley to assume a fielder's position sixty feet and six inches away and throw a tennis ball to him at batting practice speed. Wrigley would snatch the ball in his mouth and bring it back, and we repeated the process until my rotator cuff refused to rotate. After playing catch with Wrigley one day, a friend said that Wrigley was "the greatest dog that ever lived." I think I will always agree with his assessment.

During the trip with Ebbie, I brought some of Wrigley's ashes to Catalina Island, where we made a pilgrimage to the Wrigley Memorial with two members of the Catalina Conservancy. When we arrived, I opened up the film canister and stared at the crusty charcoal-like remains. I walked over to a bush that was close to a plaque honoring William Wrigley, the former owner of the Cubs after whom Wrigley Field is named. I took a few deep breaths, and let go of Wrigley.

As his ashes scattered over the plant, I flashed on the night he died in 1997. He was not quite thirteen, and had not been doing well for a couple of years. His face was pure white, and his hip dysplasia was only partially relieved with medication recommended by the UC Davis veterinary center.

Leah and I had talked about putting him to sleep, but he still seemed to enjoy life. He continued to wrestle softly with our second Airedale, Ragbrai, and on camping trips we pulled him in a wagon so he could dunk his arthritic legs in the Pacific.

After a trip to Bodega Bay, however, Wrigley developed pneumonia. He wheezed badly for several days, and two nights before our twelfth wedding anniversary things turned for the worse. Wrigley threw up all night long: first food, then phlegm, then water and then blood. I stayed up until sunrise, cleaning up his vomit and coaxing him to drink water.

At one point, a dark film started to coat his eyes. It looked like death itself, and Wrigley kept blinking it away, fighting to stay awake and to stay alive.

Around three o'clock in the morning, I walked into the bedroom where Leah and Ragbrai were sleeping and told Leah that I wanted to put Ebbie to sleep the next morning. She concurred with the decision.

When morning came, we brought Ragbrai into the living room to say goodbye to Wrigley, but she took one sniff and ran out of the room. We think that she smelled death.

I carried Wrigley into the van, and we drove to the vet. On the way, Neil Young's "Heart of Gold" came on the radio. I have never been a Neil Young fan, but when I heard that song I could not stop crying. I still get choked up whenever I hear it.

When I finally regained my composure, I lifted Wrigley into the vet's office and placed him on the steel table in one of the rooms. I knelt down and talked softly to him, staring into his eyes, as Leah stroked his side. He didn't seem scared or upset when the vet injected the poison into his veins. Within five or so seconds the pupils of his eyes dilated, as if someone had shifted the aperture on a camera. And my beloved Wrigley was gone. Leah and I stayed in the office for about thirty minutes, caressing the body of our golden boy. Then we went home and cried for several weeks.

I continued to grieve for Wrigley even after we brought Ebbie home six months later. Every now and then I would find one of Wrigley's tennis balls or look through our Wrigley photo album or hear that damn Neil Young song and just sob. I grieved more for Wrigley than for my grandmother and other family and friends who have departed, though that might seem crazy to people who don't love dogs. It seems crazy to me sometimes, until I remember how much I loved Wrigley.

If I had been by myself at the Wrigley Memorial, I might have broken down as I watched Wrigley's ashes coat the plant. I'm sure that the members of the Conservancy would have understood, but I guess my male defenses had kicked in. I simply closed my eyes for a few seconds and thought of Wrigley, and then opened my eyes and smiled at Ebbet.

I thought about Wrigley that night in our motel and cried several times. He had died five years earlier, and I couldn't understand why I still was feeling so emotional about him. I grabbed *Pet Loss* by pet loss grief counselor Eleanor Harris, one of the books I brought along on the trip. Harris argues that the grieving process can be a lengthy one, lasting several years. She also suggests that having a burial service for your pet can provide resolution, the final stage of the grieving process.

I realized that by never having a burial ceremony for Wrigley, I might have prevented the possibility of closure. I had him cremated after his death, but I never knew what to do with his ashes, most of which remain in a box on my desk at home. But when I buried a portion of his remains at the Wrigley Memorial, I finally could feel some resolution. "Resolution is a beautiful stage in the grieving process," Harris writes. "The beauty of your unique bond with your pet is that the loving relationship does not end at death. . . . Your pet's memory and love will bring you happiness for the rest of your life and then beyond" (107). I put down the book and asked Ebbie to jump on the bed with me. He refused at first, preferring to lie on the cool tiles near the bathroom. But I told him that I needed him, and he hopped up and lay on top of the covers. I cuddled him, feeling a double dose of golden love.

Over time, I have developed as strong an affection for Ebbie as I had for Wrigley, though my relationship with Ebbie is significantly different. Even though Ebbie is a Golden *Retriever*, he prefers "keep-away" to retrieving tennis balls, so I do not enjoy the playing-catch ritual that Wrigley and I shared for most of his life. But at the beach, Ebbie always retrieves his stick. I often throw it and jog in the opposite direction so he has to sprint to catch me, and then I turn around and yell "Gimme that stick," and we shift into keep-away. Sometimes I'm the chaser and sometimes I'm the chased. It is the purest form of play that I experience with Ebbie, and it happens only at the beach.

Ebbie can also fall asleep while lying on top of me, a ritual I started with Wrigley and Ebbie when they were puppies. I would lie on the floor or on the couch and pull the little fluffball on me and they would coo, fall asleep and snore. But when Wrigley was about six months old, he grew out of that habit and would jump off of me the moment I lifted my arms. But all seventy-five pounds of Ebbie continues to fall asleep and snore until I move him off of me.

Perhaps the most intimate ritual Ebbie and I share is mutual sniffing. (No, not his butt.) The practice began many years ago as a greeting ritual. I would come home and kneel down to greet him and, after licking my chin a few times, he would bury his head in my chest and start sniffing. I started to return his sniff on the top of his head while he sniffed my chest. Although my nose is much less powerful than his, I have come to love the musty, sweet, smoky scent of his fur. It is a truly sensual experience, and a powerful way of showing my affection for an amazingly affectionate creature.

In the final analysis, I truly love Ebbie, and I believe that he loves me. Do I believe that he loves me "unconditionally"? Jeffrey Moussaieff Masson might think so. He argues that dogs love "unconditionally and without ambivalence," adding: "The capacity for love in the dog is so pronounced, so developed that it is almost like another sense or another organ. It might well be termed hyperlove, and it is bestowed upon all humans who live closely with a dog" (39). In her *Companion Species Manifesto*, however, Donna Haraway disagrees, arguing that a belief in the "unconditional love" of dogs is a pernicious "neurosis of caninophiliac narcissism," especially in people who love their dogs as children. Haraway argues that respect and trust more than love are the most important aspects of a good working relationship between dogs and humans.

It doesn't matter to me whether Ebbie's love for me is unconditional or if what I interpret as love is just his need to be with a pack or his desire to get a treat or take a walk. What matters to me is that we experienced an amazing journey together and that the bond we share is a powerful one that transcends friendship and family.

POSTSCRIPT

Some key changes have occurred in my life since conducting this study. A little over a year after my trip with Ebbie, my wife Leah experienced some abdominal bloating and cramps and thought she had an intestinal problem or a bladder infection. Instead she had Stage IV ovarian cancer and was told that she might have only months to live. She had major surgery and chemotherapy that extended that early prognosis, and she died fourteen months after her diagnosis, on the day before my forty-ninth birthday.

I have no evidence that Ebbet and Hawkeye extended Leah's life, but they certainly comforted her throughout the ordeal. She cuddled Hawkeye every day on the couch and every night in bed. I even bought a bigger bed so that the three of us would have more room (Ebbie continued to sleep on the floor as usual). She continued to care for them until she was too weak to do so. One of the last things I whispered into Leah's ear before she took her last breath was that I would "take care of the boys."

The boys comforted me as well. While Leah was still alive, Ebbie and Hawkeye helped me cope with her illness. I continued to go on long walks with the dogs during Leah's rest after chemotherapy treatments to clear my head and regain my strength. They brought us important moments of laughter when they did silly things that only dogs can do, like play tug-of-war with their favorite toy squirrel for what seemed like hours.

When Leah died, they helped me to cope with a new reality and to grieve. Even at the darkest moments of despair, they kept me grounded. After all, I still had to feed them, clean them and take care of them. I started to cuddle Hawkeye on the couch and in bed. I tried to get Ebbie to join us but he still preferred sleeping on the cold wooden floors. And I continued to go camping with them, including the trip to the cove near our—now my—property in Fort Bragg where I scattered some of Leah's ashes. I will always remember and cherish the time spent with my boys after the loss of my wife.

A few years later, Ebbet died. I had felt a lump on his side and the vet said an ultrasound revealed that it was a large tumor or possibly a bone that he might have swallowed. When the vet said the worst-case scenario was that he would open him up, find an inoperable tumor and not revive him, I made the very difficult decision to put him down. I had waited too long to put Wrigley to sleep, and he had suffered from hip dysplasia and other ailments longer than I thought he should have. I didn't want to make the same mistake with Ebbet.

The vet performed an autopsy that revealed it was indeed a large tumor. Ebbet might not have survived surgery, and even if he had it would have been a very difficult recovery. He told me that I had made "the right decision," though that did not provide much comfort. My beloved Golden had died of cancer too, and I felt as though all the important members of my family were dying of cancer. I scattered Ebbet's ashes in the same cove in Fort Bragg where I scattered Leah's ashes.

I thought about getting another dog, mostly for Hawkeye, who now was home alone when I worked and in a kennel when I traveled. But I had begun to travel even more, and I just didn't think it a good idea to get another dog. I continued to take Hawkeye on walks and on camping trips up the coast, but I believed in my heart that he needed a buddy. After a few months I made another tough decision to adopt him out. It took several weeks, but with the help of the Northern California Wheaten terrier rescue, I found a match: a family with a kid and another dog. I felt very guilty about this decision at first, as I had promised Leah on her death bed that I would "take care of the boys." But I realized that "taking care" of Hawkeye meant finding him a new pack, where he could develop relationships with a mom, a dad, a kid and another dog.

I am now dogless for the first time in over twenty-five years. Not having a dog allows me to stay at the office as long as I wish, to take off on weekends without getting a dog sitter and to travel internationally for weeks at a time without paying a kennel. Although I enjoy this new freedom, I must admit that I miss having a dog, mostly because of the relational aspects discussed in this chapter. I miss having a canine buddy with whom to walk and hang out. I miss the process of getting to know a new dog and how we learn about each other's personalities. I miss training a dog and the pleasure of seeing him or her learn new tricks. I miss feeling that rare bond with a dog, regardless of how conditional that love might be. And I especially miss not having a pack, a "family," as I am now a single person living alone. I am confident that I will have a dog at some future point, and I know that I will cherish the unique relationship I will develop with that dog.

REFERENCES

Berger, Charles R., and R. J. Calabrese. "Some Explorations in Initial Interactions and Beyond: Toward a Developmental Theory of Interpersonal Communication. *Human Communication Research* 1 (1975): 98–112. Print.

Garber, Marjorie. *Dog Love.* New York: Simon & Schuster, 1996. Print.

Haraway, Donna. *The Companion Species Manifesto: Dogs, People, and Significant Otherness.* Chicago: Prickly Paradigm Press, 2003. Print.

Harris, Eleanor. *Pet Loss.* St. Paul, MN: Llewellyn, 1996. Print.

Knapp, Carolyn. *Pack of Two: The Intricate Bond between People and Dogs.* New York: Dell, 1998. Print.

Knapp, Mark L., and Anita L. Vangelisti. *Interpersonal Communication and Human Relationships.* 4th ed. Boston: Allyn and Bacon, 2000. Print.

Maslow, Abraham. *Toward a Psychology of Being.* 3rd ed. Princeton, NJ: Van Nostrand, 1999. Print.

Masson, Jeffrey Moussaieff. *Dogs Never Lie about Love: Reflections on the Emotional Worlds of Dogs.* New York: Three Rivers Press, 1997. Print.

Schutz, William. *The Interpersonal Underworld.* Palo Alto: Science and Behavior Books, 1976. Print.

Spitzberg, Brian H., and Cupach, William R., eds. *The Dark Side of Interpersonal Communication.* 2nd ed. Hillsdale, NJ: Lawrence Erlbaum Associates, 2007. Print.

8 Communicating Social Support to Grieving Clients
The Veterinarians' View

Mary Pilgram

Henry was my first child. I got him when he was two months old. We had been companions for eighteen years when he died of cancer. He was a Siamese cat that I loved dearly as a member of my family. Through his illness and ultimately his death, I became very aware of the important role my veterinarian played in communicating social support during this very difficult time. Many coworkers, friends and family did not understand my grief. My veterinarian did, and his social support was often the only thing I took comfort in.

This exploratory study examines veterinarians' perceptions of giving social support to grieving clients. Such research is important because pet loss is not legitimized by society, as pets have traditionally been viewed as property, not family members. Meyers states that "in general, animals' roles are undervalued unless animals are of direct use to people and society" (251). However, as Kenneth and Nathanial Kaufman note in their review of the literature on the animal-human bond, there is a growing body of research on the elevated status of pets in urban societies and the recognition of an animal-human bond (61). This research shows that pet owners can feel toward their companion animal as if it were a human, and consider their pet a member of their family (Butler and DeGraff 58 ; Cain 5; Hall et al. 368; Hetts and Lagoni 879 ; Sharkin and Knox 414). In addition, pet owners are willing to provide extensive medical treatment to cure or prolong the life of their ill companion animal (Kent 40). Because there are 88 million owned cats and 74 million owned dogs in the United States (APPMA) that will eventually die, there certainly is a need for this kind of research, especially if it is found that we could offer better support to those experiencing pet loss.

It is worthwhile to study the veterinarian-client relationship, as the veterinarian (or "vet") is one of the primary persons the client spends time communicating with when a pet has a terminal illness. Many pet owners cherish the relationship with their vets, and many owners feel that they can talk about their grief only with veterinary staff (Goldberg, Kerlin, Golden, and Bonica 40; Stutts 429). Losing a pet can be a significant emotional event. For some, their pet may be the only thing they love. However, not everyone regards pet loss as a significant loss (Meyers 251). Individuals who might have had an adequate social-support network following the death of a human significant other may not fare as well when the death is of an animal companion (Hall et al. 368). The veterinarian may be their sole source of social support. Therefore, this relationship is important and can impact how a client copes with treatment options and, subsequently, the eventual loss of the pet. Several of the vets in this study indicated that repeat business is guaranteed if a euthanasia procedure is handled to the client's satisfaction (being kind to the animal, being sensitive to the owners' grief, etc.). Consequently, if it is not, the client will not return and will not refer the vet to others.

Few studies have examined this specific area (Ptacek, Leonard, and McKee 366). Instead, only popular press books and articles manage to generically mention the importance of communicating with clients and "how to handle the client" when a pet dies (e.g., Boss; Gorman; Lachman 37; Lagoni and Durrance; Milani). Although the field of social support is extensively studied, looking at social support in the context of a vet-client relationship is a new application.

REVIEW OF LITERATURE

Some research has examined the grief reactions of owners upon the death of their companion animal (Smith). The grief reactions were repeatedly reported by grievers to be more painful than those associated with the loss of an immediate family member (Cusack; Quackenbush 395; Stewart 390). The grief reaction of owners can become complicated, developing into depression and suicidal ideation (Keddie 21; Rynearson 550; Quackenbush 395; Weisman 241). The research finds that in situations that involved the euthanized death of an animal, owners experienced intense guilt and ruminated about the timing of the euthanasia (Meyers 251; Stewart 390).

Grief for a Companion Animal and Lack of Social Support

It is evident that losing a pet and anticipating losing a pet are significant emotional events. However, not everyone regards pet loss as a significant loss (Meyers 251), so those grieving the loss of a pet may experience disenfranchised grief (Doka, "Introduction" 5; Doka, "Disenfranchised Grief" 223). Disenfranchised grief is defined as grief that persons experience when they incur a loss that is not or cannot be openly acknowledged, publicly mourned or socially supported (Doka, "Introduction" 5; Doka, "Disenfranchised Grief" 223). Unfortunately many people do not find the social support they need when grieving over the loss or potential loss of a pet (Baydak; Hall et al. 368; Meyers 251; Zasloff). Social support and the ability to confide in others have been shown to be important elements in weathering bereavement (Doka, "Introduction" 5; Doka, "Disenfranchised Grief" 223; Pennebaker, "Opening Up"; Pennebaker and O'Heeron 473). The adjustment process for the loss of a pet, however, may be hindered by a lack of social support and opportunities for healthy confiding in others (Beck and Katcher; Quackenbush, "The Social Context" 333; Stewart, Thrush, Paulus, and Hafner 383).

The lack of socially sanctioned mourning procedures due to cultural norms contributes to the isolation and unhappiness felt by many grieving pet owners and may also cause complicated grief. Complicated grief (Stroebe, Hansson, Schut, and Stroebe 3) is defined as a deviation from the cultural norm in the time course or intensity of specific or general symptoms of grief. For example, it may be more culturally acceptable to still be grieving the death of a parent after six months, but not the death of the family dog. This may even cause us to judge the legitimacy of our own grief reactions (Kauffman 61). Therefore, healthy grieving may not take place, putting individuals at risk for further difficulties (Lagoni et al., "The Human-Animal Bond"). For instance, James Pennebaker finds that long-term health problems are associated with an inability to confide traumatic events to others ("Opening Up"; "Traumatic experience" 82).

Social Support and Veterinarians' Supportive Communication

Social support is important to our functioning as human beings. Supportive communication has been defined as verbal and nonverbal behavior that influences how providers and recipients view themselves, their situations, the other and their relationship (Albrecht, Burleson, and Goldsmith 419). It is the primary process through which individuals coordinate their actions in support-seeking and support-giving encounters (Albrecht, Burleson, and Goldsmith 419). Carolyn Cutrona and Julie Suhr categorize social support as either (1) action-facilitating support such as performing tasks and collecting information, or (2) nurturing support which entails building self-esteem, acknowledging and expressing emotions and providing companionship (113).

Marilyn Gerwolls reports that, whereas it should be possible to rely upon veterinary professionals for support, bereaved owners may not always be able to do so, since such personnel may have difficulty dealing with their own feelings when a pet dies and may resort to icy professionalism (172). Although caring deeply about their clients (both human and animal), most veterinarians and their staff are often overwhelmed and ill-equipped to deal with individuals' emotional responses to the illness or death of a beloved pet (Lagoni et al., "The Human-Animal Bond").

An Internet search (using academic journals and databases) for vet-client communications yielded only one empirical study dealing with the topic of veterinarians' breaking bad news to clients (Ptacek, Leonard, and McKee 366), with the remaining results being popular press books and articles.

Nothing was found on vets offering support to clients. Further, an Internet search of Colleges of Veterinary Medicine with published curricula (AVMA) did not include any courses that clearly indicated the topic of vet-client communication, let alone of providing support to grieving clients. Again, this void in the research reinforces the need for closer study in this area.

PURPOSE OF STUDY

This exploratory study will focus on the relationship between the veterinarian and client and how the vets perceive that they communicate social support to clients grieving the loss or anticipated loss of a pet. Since the vet is the primary person the client will be working with to make treatment decisions, it seems reasonable that the vet should be trained to provide social support to the client during those contact times. The above literature review suggests how difficult it can be to find social support from even close family and friends, and thus, the immense value of receiving support from the vet. Therefore, two research questions were posed in this exploratory study: (1) How do veterinarians communicate social support when breaking bad news about a pet's terminal illness?; (2) How do veterinarians (and their staff) learn how to communicate social support?

METHOD

Research Participants

Ten participants (four men and six women) were recruited based on snowball sampling, which consisted of veterinarians that the researcher knew through her network. The average age of the participants was 43.14 (47.28 for men [range = 40 to 55]; 39.0 for women [range = 33 to 45]). The average years of practice was 15.85 years (22 years for men; 9.7 for women). The average number of clients that each vet sees per week was 98.75 (122 for men; 75.5 for women). All of the veterinarians were involved in small animal practices.

Procedures

A letter was sent to all potential participants introducing the purpose of the study, asking for their voluntary participation and indicating that the researcher would call to schedule an interview. Interviews were scheduled for thirty minute slots at the veterinarians' offices. Each participant was asked to complete the informed consent form, answer some demographic questions and then participate in the interview. With each vet's permission, the interviews were tape recorded. The interview questions were designed to provide information about each veterinarian's view of social support, from how they define social support to how they have learned to offer social support (e.g., training in vet school or on the job). Other questions included: How important is it to offer social support in your job?; Do you refer clients to sources of support (e.g., websites, books, counselors, support groups, pet cemetery, etc.)?; How do you know if clients desire social support from you?; How comfortable are you in offering clients social support (1 = not comfortable, 5 = highly comfortable)?; How important is the role of your staff in being comfortable offering social support? Are they given training? Please describe; Do you vary your strategy of offering social support based on your perception of how bonded your client is to their pet? Please describe.

Data Analysis

Data for analysis were derived from transcripts of ten interviews. Thematic analysis was used as an inductive approach to identify recurrent themes in the discourse (Attride-Stirling 385). Jennifer Attride-Stirling states, "thematic analyses seek to unearth the themes salient in a text at different levels, and thematic networks aim to facilitate the structuring and depiction of these themes" (387). Steps of the thematic analysis included: (1) reading the transcripts for patterns, (2) identifying themes, (3) constructing thematic networks, (4) describing and exploring the thematic networks, (5) summarizing the thematic networks and (6) interpreting patterns. In addition to the researcher, two independent coders were used to review the transcripts and formulate themes independently. The lists of themes were compared and discrepancies between any themes were discussed until agreement was reached on the appropriate themes. A theme was defined as more than one mention of an idea. Participants are referred to numerically (e.g., participant 1 through participant 10).

RESULTS

Results of this study are organized around the two research questions and themes that came from the data. Each of these themes is explored below with examples from the interview data.

Breaking Bad News

The first research question asks, "How do veterinarians communicate social support when breaking bad news about a pet's terminal illness?" The following themes emerged in participants' responses: (a) by offering emotional support, (b) by offering informational support and (c) by offering instrumental support.

Emotional Support

All ten of the vets mentioned the importance of offering emotional support to grieving clients. Positive reappraisal and acknowledging and expressing emotions were the types of emotional support that were used most often in helping a client cope.

Positive Reappraisal

Appraisal theory (Lazarus) views emotional states as resulting from the individual's interpretation of an event. Albrecht et al. indicate "that most forms of supportive communication directed at improving the affective state of a distressed individual can be viewed as suggesting or attempting to stimulate some form of reappraisal of the stressful situation" (427). Participant 3 used this strategy when telling the client that "it is OK to have your animal euthanized . . . we love our animals by euthanizing them . . . we let them go so they are not suffering . . . you've tried everything."

Participant 3 stressed the importance of telling clients that it is OK to euthanize and helping to relieve some of their guilt. She used an analogy of people living in a mental institution: "Some people can function living in a mental institution but they cannot function in reality. It's the same for a pet . . . some can't function in reality anymore and we have the gift of euthanasia in the veterinary medicine field to put them out of their suffering." Participant 4 used positive reappraisal when sitting with a client before she euthanized her pet. The vet said, "we talk about all of the good times and how the pet won't be in pain anymore and how you've done all you can for them." This vet indicated that many people feel that they are giving up on their pet if they choose to euthanize them, so offering them a different perspective sometimes helps them to consider another way of viewing the situation. Participant 9 said, "no one could've taken better care of this cat than you. Do you realize if you hadn't been the owner, he would've died a long time ago?"

Acknowledging and Expressing Emotions (the Client and the Vet)

All ten vets indicated the importance of acknowledging and expressing emotions, another type of emotional support. All ten vets indicated that they send sympathy cards to clients to acknowledge their sorrow (participants 2 and 9 stressed that their practices send only handwritten notes in order to stress the importance of the personal touch), and participants 2, 3 and 9 call the clients if they have a close relationship with them. Participant 5 indicated that she started taking one of her clients to lunch because she felt so badly for her after the loss of her dog. This participant also indicated that she shared information about her own experience with her dog's cancer and how she handled it. Participant 1 indicated that he has attended funerals for pets and has been involved in cremation ceremonies for pets.

These are ways he can acknowledge both the client's and his own emotions. Participants 1, 2 and 5 mentioned making a contribution to the Pet Trust at nearby universities in memory of the client's pet. This money is given as a scholarship to a veterinary student. Participant 4 mentioned that she provides clay paws of the pet and a sympathy card to the client, and participant 9 provides them with a cat angel pin. All ten vets mentioned that they would never rush the client after euthanasia and let them stay with their pet as long as they wanted to. Participants 4 and 8 indicated that they would ask that their clients not drive right way if they were highly emotional. They wanted the clients to know that they cared about their safety.

Informational Support

All ten vets mentioned the importance of providing informational support to clients. Each vet mentioned the importance of being clear about the prognosis of the pet's illness and laying out recommendations for treatment (or not). Participant 4 said, "we can't tell the client you need to put your animal to sleep but we can lay out recommendations and options. Still, it's the client's decision. We'll do what we can to make the animal comfortable." Participant 5 said she discusses financial constraints such as paying $5,000 to gain three to six months of life for the pet. "Every person has a different place when they are ready to let the pet go (from instantly to extreme pain)." Participant 4 mentioned the importance of being available to provide information to clients. She stated,

> We need to be there for our clients when they have questions. We need to answer their questions as fully as possible in a way that they understand. We need to be reachable when they call. I want to be able to speak with the client directly, especially if they have a sick pet, so the client can have peace of mind. We want them to feel we take the time to speak with them and that their pets are receiving quality care in a kind and compassionate way. Most people see their pets as children and they want them to be treated in such a way.

Participant 1 has a staff person with a social work background share information with clients stressing the importance of thinking about the pet's best interest. All ten participants said they always clearly explain the process of euthanasia, what to expect, what the animal might do, etc. Sharing this information, said participant 4, "helps them with their fear of this procedure. It means a lot to be with the pet until the very end."

Instrumental Support

All ten participants mentioned types of instrumental support such as the importance of scheduling appointments at the convenience of the client and also scheduling euthanasia appointments at a time of day that is less busy in the clinic so the client can have some privacy. Eight participants indicated they had a private exit for the client to leave through so they would not have to walk through a waiting room of other clients. Participants 1 and 10 mentioned that they made house calls for euthanasia, and participant 1 said he will meet clients at any hour if their pet is very ill (he met a client at his practice on Thanksgiving morning). Participant 10 indicated that she will go to clients' homes for euthanasia because "some clients don't want their pet's last moments to be at the vet because their pet just didn't like going there [to the vet] anyway."

Participants 1 and 6 have a comfort room, which is a private place for grieving clients to be with or hold their pets while they are being euthanized. Both rooms have a rocking chair and a mat you can unfold and lie down on with your pet. Both participants 1 and 6 said clients can spend as much time with their pet in the comfort room as they desire, and then the client can indicate when he or she is ready for the euthanasia procedure. Some clients even change their mind after a few hours in the comfort room, saying they cannot yet euthanize their pet. The vets indicated that they will always let the client decide when the time is right and that they will never rush the client.

Participant 3 mentioned funds to offer a client financial support if they do not have the means to provide for their ill pet's treatment. Clients who have lost pets sometimes donate to this fund (the Angel Fund) in memory of their pet so other owners can provide medical treatment for their ill pets.

Participants 3, 4, 8, 9 and 10 indicated that they purposely deal with billing issues later. They acknowledge the importance of timing; having clients pay a bill after they have just lost their pet is not good timing and could be perceived as insensitive. In cases like these, the vet says "we'll contact you in the next couple of days and discuss the details." Participant 8 indicated that her office does not put the word "euthanasia" on the bill; instead they put "balance due" to show their sensitivity to the client's situation. Participant 10 indicated that sometimes she doesn't charge for euthanasia.

Learning to Communicate Social Support

The second research question asks, "How do veterinarians (and their staff) learn to communicate social support?" The following themes emerged: (a) veterinary school training, (b) on-the-job training and personal experience, (c) personality trait, (d) need for additional training in this area and (e) training for staff.

Veterinary School Training

None of the ten participants had any coursework on communicating social support to grieving clients. All ten mentioned that they observed some vets (not all) demonstrating social support by observing in the last year of vet training in the clinical setting. All ten participants indicated that their coursework was entirely focused on the medical side of being a vet and indicated a need for coursework (including role playing) in social support. Participant 4 stated, "the compassion piece" is missing in our training.

On-the-Job Training and Personal Experience

All ten participants indicated that they really learned how to offer social support once they took their first job after vet school. They observed how other vets and staff members offered social support. Participant 10 stated, "people learn to do it (offer social support) through mentors or natural ability and others aren't good at it. There are a lot that aren't good at it." All ten participants said their own experiences in losing pets and how they were comforted was also a way they learned how to extend social support. Participant 10 said, "I have had personal experience losing pets and it's a very personal experience/loss." Participant 8 said that she learned to offer social support by "trial and error . . . was the client happy (satisfied), did they write a thank you note?"

Personality Trait (Nature vs. Nurture)

Participant 3 indicated that "you can't teach compassion. You either have it or you don't." Participants 3, 4 and 5 indicated that it takes a certain kind of personality to be a vet. They mentioned that some say they decided to be a vet because they do not like dealing with people. Participant 3 stated,

> If you don't like people you shouldn't be a vet. You have to work with people every day. You need a specific personality type to be a good vet. Clients say that "I can really tell you really like animals. I've been to Dr. 'X' and he doesn't like animals." Clients can perceive a compassionate individual versus one who is there for the money.

Participant 4 added, "vets need to have compassion for people. You have to rely on the client for what's ailing the animal." Participants 1 and 3 mentioned how competitive vet school is and that the selection criteria should involve "screening" for compassion. Both vets mentioned that they could get a feel for this from the essay the applicants have to write as well as the face-to-face interview. Applicants who are not perceived to have this skill should be weeded out. There is much more to being a good vet than having a high GPA or excellent research skills.

Need for Additional Training for Vets

All ten participants indicated the need for additional training in social support. All ten participants mentioned that more emphasis has been given to this skill in the last five years at veterinary conferences. Participants 1, 8 and 9 said, "you see more seminars on the human-animal bond now." Participant 1 indicated that "social support is the number one growth area in small animal medicine." He indicated that Kansas State's vet program is putting enormous emphasis on the area of social support now. Participant 2 stated, "we need more of it [training in social support]."

Training for Staff

All ten participants indicated that they do not provide formal training for their staff in how to offer a client social support. All of the vets mentioned that many times their staff (receptionists, technicians, etc.) spend more time with the clients than the vets do. All stated that most members of their staff are long-term employees who have developed long-term relationships with the clients and therefore have a sense of the kind of support the client might need. Participant 4 said, "my staff knows which clients would prefer a hug." All of the vets say most of their staff are pet owners and love animals and are compassionate people. "That's what draws them to this type of work."

IMPLICATIONS AND LIMITATIONS

This exploratory study demonstrates that veterinarians *perceive* themselves as communicating emotional, informational and instrumental social support to clients grieving over the loss or potential loss of a pet. In addition, the results reveal the lack of training in their veterinary schooling on how to communicate social support to grieving clients. They have either learned this on the job or through personal experience. In addition to providing training for the office staff, adding skills training in social support to veterinary curricula would be very beneficial. Training on various ways to frame messages and respond to grief may be most helpful to vets before facing the situation on the job, when the stakes are high. Some of the vets mentioned that they wish they would have had a chance to do some role playing of stressful situations in their veterinary training. This would help them anticipate some of the situations they will likely have to face.

The limitations of this study include the use of subjective data and the small sample size, so caution should be exercised in generalizing these findings. It is important to note that this study provides no evidence for the impact of the support strategies these vets offer. In addition, a more diverse sample is needed (to include rural vets, emergency vets, etc.) to clearly assess how social support is or is not communicated. Another limitation is that vets may potentially answer such research questions in a socially desirable way. Future research opportunities include interviewing clients to get their perception of how effective their vet is at offering social support and what specific messages are most helpful or unhelpful and why. The client's viewpoint will be the most important factor in such studies.

REFERENCES

Albrecht, T. L., Brant R. Burleson, and Dana Goldsmith. "Supportive Communication." *Handbook of Interpersonal Communication*. Ed. Mark L. Knapp and Gerald R. Miller. 2nd ed. Thousand Oaks: Sage, 1994. 419–449. Print.

APPMA—American Pet Products Manufacturers Association. Jan. 2008. Web. 13 Sept. 2008.

Attride-Stirling, Jennifer. "Thematic Networks: An Analytical Tool for Qualitative Research." *Qualitative Research*. Vol. 3. London: Sage, 2001. 385–405. Print.

AVMA—American Veterinary Medical Association. 2008. Web. 13 Sept. 2008.

Baydak, Martha. "Human Grief on the Death of a Pet." MAThesis. The University of Manitoba, 2000. Print.

Beck, Alan M., and Aaron Katcher. *Between Pets and People: The Importance of Animal Companionship*. New York: G. P. Putnam's Sons, 1983. Print.

Boss, Nan. *Educating Your Clients from A to Z: What to Say and How to Say It*. Colorado: AAHA, 1999. Print.

Butler, Carolyn, and P. S. DeGraff. "Helping during Pet Loss and Bereavement." *Veterinary Quarterly* 18 (1996): 58–60. Print.

Cain, Ann. "Pets as Family Members." *Marriage and Family Review* 8.3–4 (1985): 5–10. Print.

Cusack, Odean. *Pets and Mental Health*. New York: Haworth, 1988. Print.

Cutrona, Carolyn E., and Julie A. Suhr. "Social Support Communication in the Context of Marriage: An Analysis of Couples' Supportive Interactions." *Communication of Social Support: Messages, Interactions, Relationships, and Community*. Ed. Brant R. Burleson, T. L. Albrecht, and I. G. Sarason. Thousand Oaks: Sage, 1994. 113–135. Print.

Doka, Kenneth J. "Disenfranchised Grief in Historical and Cultural Perspective." *Handbook of Bereavement Research and Practice: Advances in Theory and Intervention*. Ed. M. S. Stroebe, R. O. Hansson, H. Schut, and W. Stroebe. Washington, D.C.: American Psychological Association, 2008. 223–40. Print.

———. "Introduction." *Disenfranchised Grief: New Directions, Challenges, and Strategies for Practice*. Ed. Kenneth J. Doka. Champaign Illinois: Research, 2002. 5–20. Print.

Gerwolls, Marilyn. "Adjustment to the Death of a Companion Animal." *Anthrozoos* 7 (1994): 172–187. Print.

Goldberg, A., M. Kerlin, M. Golden, and J. Bonica. "My Veterinarian Is the Best." *Cat Fancy* (2002): 40–42. Print.

Gorman, Carl. *Clients, Pets and Vets: Communication and Management*. London: Threshold, 2000. Print.

Hall, Molly J. et al. "Psychological Impact of the Animal-Human Bond in Disaster Preparedness and Response." *Journal of Psychiatric Practice* 10 (2004): 368–374. Print.

Hetts, Suzanne, and Laurel Lagoni. "The Owner of the Pet with Cancer." *Veterinary Clinics of North America: Small Animal Practice* 20 (1990): 879–896. Print.

Kauffman, Jeffrey. "The Psychology of Disenfranchised Grief: Liberation, Shame and Self-Disenfranchisement." *Disenfranchised Grief: New Directions, Challenges, and Strategies for Practice*. Ed. K. J. Doka. Champaign Illinois: Research, 2002. 61–78. Print.

Kaufman, Kenneth R., and Nathanial D. Kaufman. "And Then the Dog Died." *Death Studies* 30 (2006): 61–76. Print.

Keddie, L. "Pathological Mourning after the Death of a Domestic Pet." *British Journal of Psychiatry* 131 (1989): 21–25. Print.

Kent, D. "Euthanasia Is a Last Resort." *Veterinary Practice Management* (1977): 40–42. Print.

Lachman, Larry. "The Need to Grieve." *Cat Fancy* (2002): 37–38. Print.

Lagoni, Laurel, and Dana Durrance. *Connecting with Clients: Practical Communication Techniques for 15 Common Situations.* Colorado: AAHA, 1998. Print.

Lagoni, Laurel, Carolyn Butler, and Suzanne Hetts. *The Human-Animal Bond and Grief.* Philadelphia: W. B. Saunders, 1994. Print.

Lazarus, Richard S. *Emotion and Adaptation.* London: Oxford UP, 1991. Print.

Meyers, B. "Disenfranchised Grief and the Loss of an Animal Companion." *Disenfranchised Grief: New Directions, Challenges, and Strategies for Practice.* Ed. K. J. Doka. Champaign Illinois: Research, 2002. 251–264. Print.

Milani, Myrna. *The Art of Veterinary Practice: A Guide to Client Communication.* Philadelphia: U of Pennsylvania P, 1995. Print.

Pennebaker, James W. *Opening Up: The Healing Power of Expressing Emotion.* New York: Guilford, 1990. Print.

———. "Traumatic Experience and Psychosomatic Disease: Exploring the Roles of Behavioral Inhibition, Obsession, and Confiding." *Canadian Psychology* 26 (1985): 82–95. Print.

Pennebaker, James, and Robin O'Heeron. "Confiding in Others and Illness Rates among Spouses of Suicide and Accidental Death Victims." *Journal of Abnormal Psychology* 93 (1984): 473–476. Print.

Ptacek, J. T., Karen Leonard, and Tara L. McKee. "'I've Got Some Bad News . . . ': Veterinarians' Recollections of Communicating Bad News to Clients." *Journal of Applied Social Psychology* 34 (2004): 366–390. Print.

Quackenbush, James. E. "The Death of a Pet: How It Can Affect Owners." *Veterinary Clinics of North America: Small Animal Practice* 15 (1985): 395–402. Print.

Quackenbush, James. "The Social Context of Pet Loss." *The Animal Health Technician* 3 (1982): 333–337. Print.

Rynearson, Edward K. "Humans and Pets and Attachment." *British Journal of Psychiatry* 133 (1978): 550–555. Print.

Sharkin, Bruce S., and Donna Knox. "Pet Loss: Issues and Implications for the Psychologist." *Professional Psychology: Research and Practice* 34 (2003): 414–421. Print.

Smith, A. M. "Euthanasia of Companion Animals: The Owner's Decision Process and Grief Reaction." MA Thesis. Smith College for Social Work, 1993. Print.

Stewart, Cyrus, John Thrush, George Paulus, and Pafner Hafner. "The Elderly's Adjustment to the Loss of a Companion Animal." *Death Studies* 9 (1985): 383–393. Print.

Stewart, Mary. "Loss of a Pet—Loss of a Person: A Comparative Study of Bereavement." *New Perspectives in Our Lives with Companion Animals.* Ed. Alan M. Beck and Aaron H. Katcher. Philadelphia: U of Pennsylvania P, 1983. 390–399. Print.

Stroebe, Margaret S., Robert O. Hansson, Henk Schut, and Wolfgang Stroebe. "Bereavement Research: Contemporary Perspectives." *Handbook of Bereavement Research and Practice: Advances in Theory and Intervention.* Ed. Margaret S. Stroebe, Robert O. Hansson, Hank Schut, and Wolfgang Stroebe. Washington, D.C.: American Psychological Association, 2008. 3–25. Print.

Stutts, Judith. C. "Bereavement and the Human-Animal Bond." *Veterinary Technician* 17 (1996): 429–433. Print.

Weisman, Avery. "Bereavement and Companion Animals." *Omega* 22 (1991): 241–248. Print.

Zasloff, Ruth. "Friends, Confidants, and Companion Animals: A Study of Social Support Network Characteristics and Psychological Well-Being among Pet Owners and Nonowners ." Diss. Temple University, 1992. Print.

9   Flocking
Bird–Human Ritual Communication

Leigh A. Bernacchi

"I may know the names of the birds I see, but not the nature of my relationship to them, not how much of my own fate to read into theirs."—Jonathan Rosen

"The language of birds is very ancient, and, like other ancient modes of speech, very elliptical: little is said, but much is meant and understood."—Gilbert White

INTRODUCTION: WHY BIRDS AND WHY RITUAL

Humans long for birds. We long to be, eat, see, hear, touch, resemble and fly with birds. They represent an unparalleled freedom combined with a cross-cultural aesthetic. Throughout history, they have been our gods—the Egyptian falcon-headed Horus; lovers as in Zeus's attraction to Leda the Swan; and guides, from the raven for the Koyukuk to the raven and dove for Noah (Mowat 134–135; Bible, Genesis 8). Birders are one of the latest embodiments of this yearning—cultures of people seeking birds in their backyards and beyond for enjoyment, beauty, recreation, natural history and counting.

In the past century, nature writers have accomplished more than making birds muse and metaphor for human life; instead they have sought parallels in our existences. Terry Tempest Williams's *Refuge* describes Great Salt Lake's flooding of nesting and foraging habitat at the Bear River Refuge in Utah as a latent effect of humanity's large-scale manipulation, all too like her own matrilineal experience: Every woman in her family has been exposed to radioactivity and suffered from mastectomies. Loren Eiseley's essay "The Brown Wasps" details how pigeons seek home even when the habitat has been altered and removed under the guise of progress. Their memories, unrelenting, draw them back to the same place. For Rick Bass's youthful protagonist, a golden eagle entrapped on a hill is mother, identity and eagle. She makes the eagle's body her body. Henry David Thoreau's geese, Jack Turner's White Pelicans, Aldo Leopold's Woodcock exemplify musing internaturalist literati's vision of birds as nothing less than some miraculous kin. In fact, Jonathan Rosen's book is entirely devoted to the interconnected relationship of human reflection and bird existence:

"What has all this to do with birds? Birds say *life life life*, but something right alongside them is always whispering *death death death*. More than the blue sky, death is the backdrop against which the bird-watcher sees the bird. We go to look at them while they are still here to be seen and while we ourselves are still here to see them" (Rosen 68).

Birds *are* the proverbial canary in the coal mine. Fifteen birders and bird-watchers participated in hour-long semistructured interviews in 2009. As a birder-researcher, I have collected participant observation data at three Christmas Bird Counts, a Whooping Crane Festival and twenty birding outings with Audubon, ornithologists and private groups.[1] And although conservation was but an undertone to these events, birders expressed sensitivity for the frailty of birds, as this respondent claimed:

> They're an indicator of what we're doing in this world, just like most wildlife are. I think birds are extra sensitive to the things we're doing to our planet. We ought to be paying attention to it and educating people about it because you know these contractors go out and buy these big chunks of land and they're not thinking about wildlife. They're thinking about making money.

Discourse about birds' environmental sensitivity often centers on general ecosystem health. Birds illustrate the ways in which mobile animals respond to climate change in varying ways (Wiens et al.). Even the rhetoric of Partners in Flight, a trinational conservation alliance, emphasizes the ecosystem service birds provide as "indicators of environmental health," as well as the ecological importance of birds to all other organisms (Berlanga et al. 1). Consider too the opening paragraphs of Rachel Carson's diatribe against chemical-dependent culture, the 1962 environmentalist classic *Silent Spring*: "There was a strange stillness. The birds, for example—where had they gone? Many people spoke of them, puzzled and disturbed. . . . The few birds seen anywhere were moribund; they trembled violently and could not fly. It was a spring without voices" (20). Birds inform humans of the repercussions of their actions, but how do we learn from their actions? And where are the sites, when are the times for valuing birds for their inherency, their existence?

Some wild birds depend upon humans for resources and actions, finding creative ways to manipulate our structures for their own uses. Barn and Great Horned owls, among the most common owls, thrive in areas with agricultural fields and buildings (All About Birds). According to the Purple Martin Conservation Association, east of the Rocky Mountains in North America, Purple Martins nest only in "human-supplied" houses constructed for their colonies. Urban members of the corvid family (ravens, crows, jays, magpies, jackdaws) crack the shells from nuts with cars—in crosswalks (Klein)! However, birders and ornithologists gain knowledge of birds' capacities for coexistence through careful, patient observation as most field guides do not address the ways in which a bird affects or is affected by humans and our artifices (Schaffner). Scientists argue that many birds' range shifts are as much a response to people changing what they plant in their gardens as to climate (Weidensaul, "CBC"). Birds, perhaps as much as any nonhuman animal, call and respond, that is *communicate*, within the myriad yet myopic perceptions of humans and our actions.

For this chapter, we depart from typical birder studies, in which birds are represented in symbolic and instrumental forms. I take birding as the major text in order to pursue a ritualistic perspective on the social world of birds and humans. Rappaport, a key theorist of ritual communication, indicates that symbolic forms (mythic, artistic, etc.) lie at the depths of ritual and are the "least distinctive"; instead he proposes seeking the superficial and obvious aspects of ritual because "certain things can be expressed only in ritual" (192). This is not to say that bird-human relationships are entirely articulated through a ritual communication reading, nor is birding purely ritual with the result that all other explanations fail. Whereas symbolic life exposes a second tier of thought and response, ritual studies the primary experience, the visceral reception and production of symbols. Ritual communication, for this author and birder, provides theory and practice through which to query the cultural and communication phenomenon of birding.

Birding, I will argue in the following pages, depends upon what happens *before* our animal eyes shift from the more-than-human world[2] toward our own internal desires. What happens when human and bird meet? As I can only ask humans of their experiences and as of yet am unable to interview the birds, I represent the anthropocentric side in this review of ritual communication between birder and bird. However, I find solace in Peters's perspective that communication among the more-than-human worlds and humans is not commensurate but ephemeral. He writes at the close of *Speaking into the Air*, "If we thought of communication as the occasional touch of otherness rather than a conjunction of consciousness, we might be less restrictive in our quest for nonearthly intelligence" (256). And as recipients of messages, we could be more open to their myriad meetings. For Peters, dolphins within the agoric ocean represent the ultimate polyvocal democracy; for us, walking in the atmosphere, birds are the most outspoken and omnipresent taxa, connecting with our senses on many levels. In short, birds offer a communicative connection—one that birders understand well.

These are the stakes of studying birds and birders. Not to mention that while texts on birds continue to collect a wide interest, the act of birdwatching is America's fastest growing pastime (American Bird Conservancy). According to surveys on wildlife watching and hunting, birdwatchers compose one-fifth of the American public, numbering 47.7 million people (U.S. Department of Interior and U.S. Department of Commerce 36). This collected action of watching, hearing, calling to, pishing at, playing ipods for, naming and counting birds is a social phenomenon, learned in practice as in value—that is, a ritual.

METHODOLOGY

In order to conduct this study, I combined interviews of birders with participant observation for ethnographic fieldwork as a way of engaging and describing a culture. James Spradley has argued that participant observation is best conducted in situations where simplicity, accessibility, unobtrusiveness, permissibleness and frequently recurring activities are valued and available (52). By participating in birding activities, from field trips and festivals to personal listing and citizen science counts, I sought "to understand another way of life from the native point of view," thus learning from people how to be a birder (Spradley 3). Because birding has a strong but open social network, and access to wildlife watching areas is generally limited only by a nominal fee, I was able to participate in an unobtrusive way, often identifying myself as a social science researcher within the Department of Wildlife and Fisheries Sciences, an appropriate level of discretion for defining the human instrument, according to Lincoln and Guba.

I also conducted ethnographic interviews using a semistructured protocol (Lincoln and Guba) and digital audio recorder. Snowball sampling and attendees to local Audubon Society meetings supplied the respondent base. I was also given two additional interviews completed two years prior by a colleague, totaling fifteen formal interviews. To increase the scope of birding culture I have also included texts on how to bird; field guides; first-person narratives in book and blog forms; bird counts; histories on the subject; and environmental literature. In short, the ethnographic field notes I have collected from 2008–2012 include fifteen interviews, hundreds of hours of participant observation, first-person narrative texts and secondary materials from private, public and nonprofit organization sources. In the end, I have become a birder by process of interacting and learning from the texts and conversations from which anyone could learn how to bird.

I feel it is important for the reader to understand that although I never intended to become what I study, I am a birder. Like the author of *The Big Year*, "Slowly but certainly I realized I wasn't just pursuing stories about birdwatchers. I was pursuing the birds, too. . . . I needed to see and conquer" (Obmascik xi). Whereas I do not feel the need to conquer, I am interested in collecting birding experiences, and have been involved in some great birding rituals. I had been warned that "raptors are a gateway drug," and I kept watching anyway. Perhaps this is what happens to humans when we participate in animal-human communication: We want to talk and to listen and to participate more.

RITUAL COMMUNICATION THEORY

Whether it is Emile Durkheim's serious life, Mary Douglas's purity and danger or Mircea Eliade's sacred and profane, the binary of ritual and not-ritual distinguishes between regular life and religious life. This is not religion in the typical connotation we use in daily conversation but the ways in which the world adopts meaning. For Eliade, the context is less about choosing a religion; he denies the wholly existential agency of the modernist: "we are confronted by the same mysterious act—the manifestation of something of a wholly different order, a reality that does not belong to our world, in objects that are an integral part of our natural 'profane' world," but instead arrive, alight, observe, fly away of their own volition and "something sacred shows itself to us" (Eliade 11).

However, the ability to experience the manifestation, to access the sacred meaning, depends on a transition through experience: "the initiate, he who has experienced the mysteries, is *he who knows*" (Eliade 189). Initiation requires people who have had this experience to structure the performance, to transmit information *and* impart knowledge and significance to the information. The participants in ritual are hierarchically defined, although these hierarchies may transform in the liminal space, and those in positions of privilege (experience and knowledge) perpetuate coded and defined ways of being (Rappaport; Douglas). For Rappaport:

> The point to be made here is that this relationship of the act of performance to that which is being performed—that it brings it into being—cannot help but specify as well the relationship of the performer to that which he is performing. He is not merely transmitting messages he finds encoded in the liturgy. He is participating in—becoming part of—the order to which his own body and breath give life. (Rappaport 192)

For example, in birding, the experience means little until someone else can recodify the value of the ordinary object into a sacred subject. Through training with other birders or bird texts, the Socratic discussion transforms the open participant.

Novice: What does it mean to see a bird you have never seen before?
Expert: Nothing. Except that this is now socially, cumulatively, personally, ecologically and scientifically significant.
Novice: What does it mean to see the yellow-crown feathers of a Myrtle Warbler?
Expert: Nothing. Except that it is rare and you have been close. You may not see that for a while, if again.

Another way birds impart the significance of their sighting by humans is through this interaction. Many novice birders cite the first time they truly saw a bird; they recall the species, place and age when this occurred; and the bird-human interaction led to value changes. Once a participant chooses to engage in the ritual and has learned some of the ritual of birding's structures, there is an in-between place experienced when the bird arrives and none of the birders have identified it. This is a rite of transition, where life is different upon seeing *and* identifying the bird (Turner; Van Gannep). The space through which the birders passed is known as the liminal space, a threshold where everyone is relinquished from their roles and still engaged (Rothenbuhler).

Ritual communication is uncontrollable yet calculated and structured. It depends upon self-perpetuation, symbols connected, configured and mutually supportive, yet flexible enough in order to change, be co-opted and contrived for other situations, creating a shared world of meanings among individuals. However the "liturgy" of birding may seem, it is entirely dependent upon the manifestation of birds. Should there be no birds, there is no experience. Further nuance within the reactive ritual depends on which birds appear, and how and where, as the values of sacred objects are disparate and uncertain depending upon place and case.

Lest it seem I am imposing circular logic to define the ritual by the ritual being defined, I would like to identify the key, "obvious" aspects of ritual, although there are as many definitions as authors. The synthetic definitions of Roy Rappaport and Eric Rothenbuhler provide succinct and lucid criteria to apply not only to birding but to other animal-human interactions, and perhaps beyond to those of the more-than-animal world. Levy discusses some of the nuances that distinguish "ceremony" from ritual for Rappaport and asks, "When is a ritual not a ritual and why do some parts of a 'ritual' . . . seem more like a 'ritual' . . . than others?" (154). I find Rappaport's succinctness useful: "the formal, stereotyped aspect of all events" consists of formality, performance, seriousness, voluntary action, repetition and congregation (Rappaport qtd. in Levy 154).

Formality implies a structure or pattern, where certain actions are accomplished at the appropriate time with the proper conduct. The symbols, embodied in persons and in objects, have material effects. There is a tension between structure and mutability, however, because performance affords room for the individual response. Therefore, rituals are embodied, not enacted; spontaneous, not scripted. For Durkheim, serious life is action regarding the sacred. The sacred subject of the bird, manifesting itself and being regarded, can theoretically be supported by the invaluable aspect of communication with *the other*, that which is not human produced (Peters). Ritual also subscribes to what is beyond value instrumentally or rationally while simultaneously denying frivolity or recreation. A person must participate of his or her own accord, at repeated or circular intervals, for involvement in a social action. The individual who performs within a ritual is always doing so with respect to a collective. In short, I define ritual as *a social process of (dis/re)ordering through interaction with the more-than-human world, where common objects manifest themselves within a system of existing or altered values to become sacred objects.*

Birding performs ritual. When the aesthetic and inexplicable experience undermines presumptions of both recreation and scientific instrumentality, rationalism swirls with emotions of hope, disappointment and excitement in experiencing a bird. It is a rare, and therefore special, time: One cannot bird all the time every day for the same birds—in their own circadian rhythms, their personal phenologies and annual migrations, their molts and matings, the lives of birds permit only intermittent yet repetitive interaction. Roger Tory Peterson's son said, "My father used the comings and goings of birds as both a biological clock of sorts and a litmus test for the condition of the environment" (Peterson ix). Finally, even if birding is a solo affair, a dialogic view of natural history and evolutionary interaction would indicate that the social process informs how one 'birds.' Natural history is a process of generalization and specification through observation over time and repeatedly, itself dependent upon ritualistic attention to the more-than-human world, both the uncategorizeable and classified. Social action suggests a structure through which the performance of interacting with birds can be embodied by and transformative for the birder. The following descriptions from the case study will enrich our understanding of bird-human relationships from a ritual communication perspective and provide necessary sites for discussion of conservation ethics.

RITUAL COMMUNICATION ASPECTS OF BIRDING

Birding, semantically, relies on the noun—a collection (taxa) of feathered animals—to create a dependent relationship. Just as skiing depends upon skis, birding depends entirely upon birds. The noun transfers into a gerund form ("Don't tell mom, but I'm birding again") and creates a new noun form ("Let's go birding"). Because of this transfer, the noun "birding" can be broken up into the imperative, as in, "Bird the Gnar!"—a carpe diem equivalent from Steven Tucker, author of the blog "Bourbon, Bastards, Birds." To take the present indicative as personal statement, one could utter, "I bird," bringing very close the subject and the referent, although this is less common in spoken vernacular. More common is to adopt the general name of the referent: I am a birder. An interview respondent expressed a preference for identifying with neither birder nor birdwatcher, but instead simply with the verb in action: "I watch birds." This phrase avoids some of the pejorative or "pigeonholed" connotations of birdwatchers: the "Jane Hathaway" character from *The Beverly Hillbillies* in a many-pocketed outfit; according to a respondent, "some little old lady in tennis shoes and a floppy hat."

Serious bird aficionados I interviewed put a finer point on the exclusive-inclusive dynamic of terms: "So birdwatcher is a mildly, amusingly contemptuous term. Birder is something that the birders use, other people usually don't know." On the extreme end of the artificial continuum of birdwatchers and birders, birders list their sightings and keep records, pursue vagrants and rarities and keep abreast of others' sightings through call chains and email lists. On the other end, birdwatchers may never leave the comfort of their backyard, porch or house; feed common species (including squirrels) alike; and often do not make records beyond their memory. A respondent noted of university campus birders, "It's been part of this community whose boundaries are unknown. I'm always finding more people." Social bonds are formed in proportion to the intensity of an activity—think of the differences in social bonds between elite mountain climbing partners versus novice personal backyard gardeners. This partially explains why the social interaction of birdwatchers, on campus or beyond, is not as strong as that of birders. Nor do they exhibit mutually recognizable behaviors in public spaces. One Utah birdwatcher has refused to learn the names of birds in order to preserve his wondrous experience of them. And Texas master naturalist and professor of English Jimmie Killingsworth has decided that, at some points, "I want my birding to be something easy and profoundly useless" (591). One more collective explanation from birding author Mark Obamscik:

> Birding is hunting without killing, preying without punishing, and collecting without clogging your home. Take a field guide into the woods and you're more than a hiker. You're a detective on a backcountry beat, tracking the latest suspect from Mexico, Antarctica, or even the Bronx. Spend enough time sloshing through swamps or scaling summits or shuffling through beach sand and you inevitably face a tough question: Am I a grown up birder or just another kid on a treasure hunt? (xii)

The two terms "birder" and "birdwatcher" represent a dynamic relationship and important distinctions within the cultures, but compared to humans who do not interact with birds, they are similar by definition, warranting a single term. Birdwatcher and birder connect through the rite of transition—the bird manifesting itself in the same place as the person, the person experiencing the bird. For the human, at least, something is different afterwards. Phenomenologically, the birder depends upon an experience of the ears, eyes, body and memory (Schaffner 5). In this chapter, in part because of the fluid definition provided by respondents and for practicality, I use the term "birder" to indicate someone who explicitly seeks birds and participates in noting natural history or names.

A DAY IN THE LIFE OF A BIRDER

In order to explain birding culture, I invite you to join a group on a typical field trip.

You rise before the sun, meet at the designated spot, caravan in groups of 3–4 in SUVs, sedans and vans to known wildlife watching. In the car, you exchange light conversation and coffee. Upon arrival, just as the sun is breaking, you bumble your cup and all the trappings of birding: occasionally a cigarette (depending on the crowd), bird book (Sibley's, a monstrous tome of paintings superior to photos for most birders in their applicability to field specimens), scope, tripod, binoculars with or without crossback straps, hat. Pen and paper, or write-in-the-rain yellow notebook and pencils in your shirt pocket. Maybe you have even noted the cloud cover, temperature and time and prepared columns for the four-letter codes for bird-name shorthand, very ornithological. You have a bird checklist for the area. There's wind. There are fire ants. There's your iPod queued up for a potential species in question and speakers in the bag too. There are patches and pins to show where you've been. There's name brand gear and faded birding festival shirts. There are acquaintances from around the region.

And there's a bird.

Liminal Space.

You see: it's on a fencepost, a bird singing and singing. You throw your glass on it, focus your binoculars. You vocalize its presence to alert others, connecting them to the bird. You try to remember the song. You search patches of color, the shape and size relative to the other birds. You look at the way it moves behind the brush and back to the post. You search your memory for similar experiences. You lose the bird. You find it again. Who is it? You render hypotheses. Wingbars. Eyering. Tail blunt. Bill relative to head. Bill sharpening against branch. What about the coverts?

At this point the bird is not yet a sacred object for ordering the place. When the bird is unidentified yet present. When a group of birders is in the process of identification, liminally hanging between the potential experiences, free from the set knowledge of species and dwelling in the mystery of not knowing the bird. The birders are experiencing communitas.[3]

Someone states the species name. The game is over. Some people watch the bird longer. Some mark it on a list. You reminisce to the last time you saw the bird. After some time, hours or minutes really do not matter, in the elements, you return to the car, piece through the book, request assistance, deliberate with the group. Not sure. Some lunch, more drive-by birding, people put their personal lists online or nowhere.

Identification

Birding as a ritual centers on identifying the species of the bird. It is difficult to attribute natural history knowledge or characterize the individual if it has not been named to the taxonomic level of a species: standard protocol for human-biology interaction (Yoon). The more-than-human world enters human understanding through language as a discrete object represented by a single word title (e.g., Veery) or set of words as an epithet (e.g., Yellow-bellied Flycatcher). Often names refer to larger groups such as genus, relating the species to other morphologically similar birds, but as genetics research adds more details, historical titles persist in the halls of the meetings of ornithological societies. To this end, birders speak a general language to name birds by their kind. The standard of the name supports the standard of identifying.

Throughout the history of birding, identification of birds has been limited to the species, but age and sex are becoming increasingly important to advanced birdwatchers. One individual planned to devote the spring to learning the subtle differences between male Northern Cardinals born that year and females. Similarly, another respondent was revered by others in the group for being able to distinguish ages of gulls. The question of whether the gull is more valued because of this enduring inquiry into its age is spurious, but following from the maximization-of-liminal-space argument I am attempting, the more interest in ritual to identify the present bird, the more engagement with the bird, and perhaps, the more *sacred* its appearance. Species identification reigns primary, but noting behavior and other characteristics enriches the relationship with the birds and shares more of their personal life with birders. The bird is no longer an object or image, but an entity all its own. It is important to note that this is part of the ritual for more advanced birders. On the other hand, backyard birdwatchers, often debased by advanced or well-traveled birders for their focus on a small set of local species, have the great opportunity to encounter individuals repeatedly, as one of the respondent rejoiced. And birders who scope nesting sites, online[4] or in person, can often develop affection for (a) particular bird(s).

Early bird identification field guides, although based on "collected" specimens, replaced the previous method of hunting, where 'one in the hand is better than two in the bush,' by focusing on the aspects of the bird that are distinct and visible to the naked eye, opera glass or best available technology. The organizer of the first Christmas Bird Count, Frank Chapman, who famously identified hundreds of individuals and parts on the fashionable hats of women in turn-of-the-century New York City avenues, preached that "you learned 'to recognize bird friends as you do human ones—-by experience'" (Dunlap 185; Price). This pattern continued through later field guides, the most important technology a birder carries according to field guide historian Thomas Dunlap,[5] where readers were educated to pay attention to the general impression of the bird, the gestalt of the species and the "unchanging physical features of a bird" with respect to what makes sense for the geography and the behavior. They compared humans to birds in that "you could put a name to [your] 'spouse or best friend a block away,' a distance too great to allow recognition of diagnostic features" (185). One respondent explained how he recognizes birds:

> I'll pick squirrels out in the woods way in the distance just because I'll see an unusual movement. And I'll see it's a not natural move. I can tell

there's a bird there. A flash of color or a silhouette even. I can tell a lot of birds by their silhouette. You know at dusk, when it's almost dark, and you really can't see them. You can tell a flock of pelicans from a flock of cranes from a flock of geese just by their silhouette and their flight patterns.

In fact, illustrator and author David Allen Sibley suggests that we leave the book behind, pay attention: "The first rule is simple: *look at the bird*. Don't fumble with the book, because by the time you find the right picture the bird will most likely be gone. Start by looking at the bird's bill and facial markings. Watch what the bird does, watch it fly away, and only then try to find it in your book" (qtd. in Dunlap 10). This would not only improve the ability to take field notes and ostensibly memorize the bird and bird experience but extend the liminal space created through unknowing identification.

Novelty

Bird guides matter-of-factly state that if you are not birding you cannot possibly identify birds. According to Bob the Birdman's blog, "I'm never not birding." Many rarities are found in the most common and abundantly birded areas. During a birding festival in California, a renowned birder discovered a vagrant in the trees above the whole congregation at the closing events. His friend mused, "All these birds are there, just ready to be seen."

The point at which there is the least amount of hierarchy occurs when one or many people have seen a bird but none have identified it (the primary response). It is the time where anyone with the right combination of past experiences could respond. Some choose not to respond to the question immediately, though, because of the imposed value of accuracy within the group. Not all birds are equally seen and not all sightings are equally accepted. While Ken Kaufman instructs birders to question authority, experts control historical and regional data, festivals, field trips, events and, in sum, the ritual.

Birders recognize the religious aspects of birding through their comment and practice. In one interview, the respondent denied the power of "self-appointed gurus" and "a self-perpetuating, self-appointed priesthood." He dismissed their regulatory power over the data sets because it is all "pseudo-scientific." Yet those records make it into regional maps. One respondent, when just a novice, was responsible for contributing to the first sighting in the county and adding "a little dot." The state rare bird committees, ornithological societies and online databases such as eBird are the end of the road in many ways. Novices must have more proof than others to go against the convention of previous sightings, much like a supreme court. Their perspective is salient because it establishes and maintains a status quo of scientific rigor, although it may not be about science at all, a value

of accuracy for birding. For those who are outside of birding, the seriousness may seem to be taken too far. Responses have been to paint birders in
a negative light, as obsessive at best. Popular representations like the Steve
Martin film *The Big Year* question the worth of paying so much and leaving
family and friends to be the person to see the most birds in North America
in a year.

Listing

The list is a private form of discovery. Once a bird has been identified in
terms of species and sometimes sex, then myriad responses are possible, but
until that identification occurs, everyone is occupying a space of not knowing, but knowing that they may eventually know: the liminal space. For
each new sighting or call or song, the liminal space is regenerated, and the
birders respond to this manifestation. A day of birding is only as sacred as
the number of birds accounted for; the list is a representation of the number
of liminal spaces occupied, the number of times the sacred manifests in the
face of the profane.

Accuracy is the most important virtue of listing for most birders. It is
important for birders to be correct in the eyes of their peers, realistic within
the true ranges of species and for their own records and personal fastidious-
ness. Some birders will not add the identifications of birds made by others to
their list unless they have verified those identifications with their own eyes.
One commented on the difficulty of discerning another birder's skill in a
group, which partially explains the carefulness. I observed two advanced
birders keep different lists for the same field trip, birding almost the entire
time with each other. When asked about this, they stated that they wanted
to keep personal lists and would not later combine them for submission to a
data site. Accuracy in identification and listing is valued by different birders
for different reasons and cannot be explained entirely by the ritual of listing. Through other facets of listing we can explore additional values.

One birder transferred ten years of data from paper notes to the Cornell Lab of Ornithology database and citizen science resource, eBird. She
admitted about listing,

> It's a complete obsession. I buy every checklist. I have several guides for
> Arizona and Texas. I'm going to go to Florida in a couple of months
> and I've already bought three volumes. I usually input about 4–5 lists
> per month. I'll tally my backyard. And one to three, depending on how
> many counties we go through, for each field trip on eBird.

Her experience shows some of the ways in which birders distinguish
the conditions and experience of interacting with birds: by number and
quality and among geographic and politically defined regions (yard,
fifteen-mile-diameter circle, county, state, country, life). There are no

rules, as some birders mark simply in the back of their field guides or keep small notebooks. Uploading data to larger sites creates a collective list for a region.

The greatest adventure for a birder is seeing a "life bird"—a bird the individual has never before seen. Also known as a seeing a "lifer," the experience has been compared to economics: "the equivalent of reaching into your pockets for change and coming out with the Hope Diamond"; to sports: "It depends, like many males in the U.S., I am defined by achievement in some sense. I'm very interested in seeing something I haven't seen before. It's an adrenaline rush. Like everything has come together"; and to other rites of passage: "How'd you feel the first time you were in the backseat of a '57 Chevy? You know, your heart racin' and your blood pressure up and at my age that is about as big of a thrill as the backseat of the '57 Chevy was." Yet others compared life birds to more casual experiences:

> Yes, it's a good feeling. It's like making a good play in softball. It's not quite as good as getting a paper accepted for publication. Maybe asking a particularly clever question at a scientific conference or something. It's pretty high up there on the good feeling thing. And because it becomes rare and rarer. . . . But it dies down after a day or two.

An ornithologist I witnessed in the Amazon oozed excitement for wildlife all day only to render taciturn statements upon seeing one of the only birds he had not seen in the region. As Van Gannep writes in *The Rites of Passage*, "the first pregnancy and first childbirth are ritually most important" (175); so too it could be with life birds. There is a time after people have recognized they are collecting life birds and before the law of diminishing returns devalues each new bird, after which they are rare experiences and the serious education that experts must impart to learners overshadows their personal satisfaction.

The antidote to bird listing ennui is "the big year," perhaps the most publicized aspect of birding and one that the fewest people have the ways and means to achieve. However, it has greatly affected the importance of the list in birding. When Kenn Kaufman broke from an affluent American elderly male birding tradition and hitchhiked around the continent for his big year, "the list was the thing" (Dunlap 169). And seeing as many birds, listing as many birds as possible within the U.S. political boundary within the distinct time period of January 1 to December 31, is the high holiday of birding ritual. Attested simply by the honor system, the big year birder is followed and celebrated, as well as envied, by birders who watch the tally of their lists climb on their blogs or personal email lists. However, competition, as I will discuss later, often causes their list to be secret until the end of the year. According to the American Birding Association, the current record for North America stands at 745 species.

Cooperation

"Did y'all get that?" This was a phrase that I initially criticized for taking ownership of the bird and emphasizing the listing of each species instead of the enjoyment. But upon closer and continued observation, it is really an opportunity for people to share sightings efficiently and expediently. If someone has not seen the species, then others will often share their spotting scope or provide accurate directions for finding the bird. Guides will check to make sure everyone sees a bird, but group members also check among their friends and those close to them. Some highly skilled birders will find all of the species in the area first and then slowly release the number so as not to overwhelm the group. During counts, the focus is not on seeing the birds, and often the recorders of sightings see very few birds themselves.

"How many did you get?" This is an entirely different question and centers on competition, but in the case of Christmas Bird Counts, the high holiday of birding, the focus is on the total group contribution to a circle, completed with a dinner for all the participants. On regular birding field trips, though, if birders do not see the bird personally, they are less likely to add it to their daily or life list. In one case, when a man could not identify a bird at a great distance but was accompanied by someone who said it was a certain species, he did not add it to his list because he did not know if the identifier was more or less skilled than himself.

David Scott, a birder and recreation theorist, has written that the more skilled individual birders become, the greater their contribution to the group's ability to learn and teach. He explains that the law of diminishing returns affects serious leisure and minimizes personal satisfaction. As one respondent noted:

> I lead trips. I never, virtually never, go out birding by myself. Somehow my time budget says that if I'm doing this it's got to be with somebody; I've got to be sharing my expertise. I have to be using my time efficiently. I don't do it as a relaxation, recreation thing. In fact, birding for me now is almost always something where I try to teach somebody else something . . . I've seen every bird, virtually, that you could see in the United States.

Competition

There are many aspects of modern birding that perplex the general public, but the strongest reactions are to rabid birders' sense of ownership and competition. Late-night talk show host Conan O'Brian satirized the competitive undertones in birding when he joined the New York City Audubon on a bird walk, "This is a birding group here. This is another birding group here. So it's about to get pretty nasty." He proceeded to egg on the two representatives in a comparison of their bird experiences for the day: "I saw a male summer tanager. Heard it and saw it." "I saw a female summer tanager." As one respondent explained:

> There is a group that I would call the more or less normal people. . . . People that you wouldn't automatically single out as being birders. Then there are the rabid birders, usually young aggressive males whose testosterone for some peculiar reason has been channeled into birding, and I suppose there are a few females like that, but mostly guys. They go out and do it in a very competitive way. They can be amusing or they can be annoying.

Competition is one means of creating hierarchy within the birding social world, which affords opportunities for ritual "high priests" to lead others. Competition also partially helps to maintain the key value of accuracy in data collection from birding experiences and therefore data points. It also reduces the experience of dwelling in the liminal space of not knowing the bird, because experts can very quickly identify birds. There are many stories of people who travel great distances to see a bird only to add it to their list without watching it, thus ticking the collected experience off of their life list. Intrapersonal and interpersonal competition drives this sport and some of the scientific aspects of it, but is not the sole determinant in birding.

In fact, competition among birders may distract from the competition inherent in survival. For instance, there is a darker undertone to accuracy-driven lists that pervades ornithological studies, establishing dichotomies among teams. The Christmas Bird Count (CBC), conducted by citizens, is subordinated to wildlife professionals' Breeding Bird Survey (BBS) data. Meanwhile, the Great Backyard Bird Count is subject to greater scrutiny by experts because backyard birders are not as adventurous or perceived to be as skilled (even though the data cycles through the same online site, eBird by Cornell Lab of Ornithology); its range is smaller and its data collection increments are much shorter than those of the Christmas Bird Count, yet many who participate in CBCs also contribute to the Great Backyard Bird Count. Even North American ornithologists working on the CBC/BBS data assimilation suggest that these data collections are superior to European studies: "In North America we have information over a much broader area, and we don't have the same uniformity of climate change as in Europe. . . . So we can tell a more complex story" (qtd. in Weidensaul, "CBC" 13). Yet bird conservation will be the primary issue for birders as they experience greater changes, not all due to novelty in new places.

CONCLUSION: CONSERVATION AND THE RITUAL RELATIONSHIP

Citizen science has been touted as the next wave in "big data" collection and science. The data sets are massive and accurate, and the process of participating and contributing fosters individual understandings and appreciation for science. In short, it is the Deweyan response to technocratic governance. It has also become part of the ritual of birding, both cooperative and competitive. Yet the theory of ritual communication assumes no direct link between rational collection of information and improved action, in this case conservation, begging the questions: How does the participation of birders and birds in ritual lead to better conservation and not just better information? How are ethical relationships emerging among birders and scientists? I wonder if augmenting the rationalist propensity of birders encourages birders to pursue identifications and data points in lieu of lingering relationships with birds or a bird as an individual. This changes the ritual of watching the bird into one of pursuit, always at the root of this natural history avocation (Weidensaul, *Of a Feather*).

Birders have noted changes to species ranges because of their complex record keeping. For instance, one man noted that there were never White-winged doves in North Texas forty years ago but that they now represent the dominant species. Others have commented on phenological shifts because some keep records of the first time they see the bird in that area each year. They can track the timing of arrivals and departures, and often submit to larger data sets like eBird that can create a much broader picture of bird response to land use and climate change. Whereas I have little sense of how many people actually create these range maps from the larger data sets, it seems that birders value their own records. These records provide information about birding opportunities and form an annual and seasonal map for birding ritual.

Birds find suitable habitat in many of the places humans have modified for their uses: urban and suburban parks, barns, bridges, agricultural fields and sewage treatment ponds. These sites have the potential to be elevated from their sullied position of "human-made" and therefore not-wilderness in the eyes of birders, a problem William Cronon identified as central to the improvement of the environmental movement's conceptualization of what qualifies as Nature. From a ritual communication perspective, the birds as sacred objects within a rite of transition serve a greater function of ordering an otherwise sordid landscape: A place of multiple human uses like a sewage pond is also used by something other, an animal who has manifested itself there, lives there and survives there—a Palm Warbler after the Christmas Bird Count on a California sewage treatment pond bank. Dozens of birders, having birded since six in the morning, came to the area and marveled at the nontraditional scenic beauty, the reflection of the yellow bird in water sullied and cleaned for continued use by humans—one more bird. In Mary Douglas's terms, the disordered landscape has the potential to be reordered as valuable to birders:

> Granted that disorder spoils pattern, it also provides the material of pattern. Order implies restriction; from all possible materials, a limited selection has been made and from all possible relations a limited set has been used. So disorder by implication is unlimited, no pattern has been realized in it, but its potential for patterning is indefinite. (Douglas 117)

Birders' pursuit of wild birds in strange places has the potential to guide people to see beyond the tenuously perched dichotomy of wholly human spaces and a separate and monolithic "Nature" and instead reenvision in a more complex, postmodern ecological manner. After all, "birding grew from a common human fascination that was directed by the culture's accepted authority about the world—science—and was driven by a desire to keep and keep touch with a nature while living in a society that threatened it" (Dunlap 199). But because birders focus on wild birds, they have a different organization of the more-than-human world than even ecologists, ornithologists or conservationists, with whom they align in other ways: Their science is one of an interest in seeing birds regardless of where they fit. Aside from "trash birds" or "bad birds,"[6] the rigorous division of boundary maintenance between native and nonnative, worthy of conservation and unworthy of conservation, is not of interest to birders as it is to conservationists (Milton). Here novelty, like biodiversity, provides a new vision for conservation action, one that might better deal with present and unprecedented challenges, such as species migration to new areas due to rapid climate change.[7]

In a more general sense, Rothenbuhler has written that the greatest benefits of ritual are the lack of coercive effects on social action, the opportunity to show what could be possible, if only temporarily (15). Similarly, birder and historian Scott Weidensaul has pointed to the "twin polarities of modern birding": birds as marks on a list or birds as living, wild creatures full of their own inherent value (281). And by returning to communication scholarship, we remember that "empathy with the inhuman is the moral and aesthetic lesson that might replace our urgent longing for communication" (Peters 246). A ritual reading of birding potentially leads to a renewal of clarity in the bird-birder relationship, which is based on what a respondent clearly stated as the reason for participating in this activity at all: "It's the birds. That's what it's all about. That's why I do it. That is what it is all about. It's the birds."

In a poignant moment for conservation, David Attenborough's *The Life of Birds* documentary series shows the forest songs of the Lyre Bird, a master of vocal mimicry, singing the songs of all that surrounds it: local birds, car alarms, camera shutters, chain saws. Other birds may be singing or communicating in similar ways, telling us more about our relationships with the more-than-human world than we thought possible. What are we saying in return?

NOTES

1. Research conducted under Internal Review Board institution (Texas A&M University) number 2010–0355.
2. David Abram employs this term to indicate "nature" or "environment" while subordinating anthropocentrism. I have adopted it here in order to maintain

this hierarchy and include humans within the world of relationships with all beings.

3. "*Communitas* is the domain of equality," the feeling of belonging to a community or group of commonality and connectedness in which all involved are (nearly) equal (Mangueira qtd. in Turner 132). The connections are more to the group than to the individuals within the group, and the group can take many forms, ranging from strong bonds of kinship to a bridge club (Turner 45). In Victor Turner's canonical analysis of ritual in "Carnival in Rio," he writes:

> If "flowing"—*communitas* is "shared flow"—denotes the holistic sensation when we act with total involvement, when action and awareness are one, (one ceases to flow if one becomes aware that one is doing it), then, just as a river needs a bed and banks to flow, so do people need framing and structural rules to do their kind of flowing. But here the rules crystallize out of the flow rather than being imposed on it from without. (133)

What I mean by *communitas* in birding is that all birders are equal as they have seen the bird but not identified it yet. Although experts can identify almost immediately a bird in flight or in the open, they can still be stumped by vagrants, so *communitas* is possible in many birding situations. In birding, opportunities for equality persist despite expertise because the manifestation of the bird is ephemeral, as with the group and the structure within the group.

4. We can never know what effect observing nestlings through technology has on their behavior. For an introductory discussion see Rinfret, and for critical discussion with particular attention to zoos and the hegemony of the ocular see Deluca and Slawter-Volkening. These inquiries into technological, on-demand views (essentially "eco-pornography") of nonhuman animals provide one path for dismantling speciesist rhetorics.

5. In his book *In the Field, among the Feathered,* Thomas Dunlap writes:

> Birding's greatest contribution to American life was to give people an activity that involved them in nature and to give them a voice in preserving it. Field guides played crucial roles. They initiated people into the community and instructed them in its ways and lore. In text and pictures they said what was important, told how to practice the craft, even what to call the birds. In the field they served, as much as binoculars, as a member's badge and an introduction. (204)

6. See Pat Munday, this volume, for a discussion of the negative characterization of ravens.

7. See Weidensaul, "CBC," and Wiens et al. for further discussion of climate and bird conservation.

REFERENCES

Abram, David. *The Spell of the Sensuous: Perception and Language in a More-Than-Human World.* New York: Pantheon Books, 1996. Print.

All About Birds. Cornell Lab of Ornithology. 2011. Web. 26 May 2012.

American Birding Association Blog. American Birding Association, 2011. Web. 26 May 2012.

Attenborough, David, perf. *The Life of Birds.* BBC Video, 2002. Film.

Berlanga, Humberto, Judith A. Kennedy, Terrell D. Rich, Maria del Coro Arizmendi, Carol.J. Beardmore, Peter J. Blancher, Gregory S. Butcher, Andrew R. Couturier, Ashley A. Dayer, Dean.W. Demarest, Wendy E. Easton, Mary

Gustafson, Eduardo E. Iñigo-Elias, Elizabeth A. Krebs, Arvind O. Panjabi, Vicente Rodriguez Contreras, Kenneth V. Rosenberg, Janet M. Ruth, Eduardo Santana Castellón, Rosa Ma. Vidal, and Tom Will. "Saving Our Shared Birds: Partners in Flight Tri-National Vision for Landbird Conservation." Ithaca, NY: Cornell Lab of Ornithology, 2010. Print.

"A Brief History of the Purple Martin and the Purple Martin Conservation Association." *Purple Martin Conservation Association.* N.d. Web. 26 May 2012.

Carey, James W. *Communication as Culture: Essays on Media and Society.* Rev. ed. New York: Routledge, 2009. Print.

Carson, Rachel L. *Silent Spring.* Rev. ed. New York: First Mariner Books, 2002. Print.

"Conservation through Birding." *American Bird Conservancy.* 2010. Web. 26 May 2012.

Cronon, William. *Uncommon Ground: Toward Reinventing Nature.* New York: W. W. Norton & Co., 1995. Print.

Deluca, Kevin M., and Lisa Slawter-Volkening. "Memories of the Tropics in Industrial Jungles: Constructing Nature, Contesting Nature." *Environmental Communication* 3.1 (2009): 1–24. Print.

Douglas, Mary. *Purity and Danger: An Analysis of the Concept of Pollution and Taboo.* Routledge Classics. London and New York: Routledge, 2005. Print.

Dunlap, Thomas R. *In the Field, among the Feathered: A History of Birders & Their Guides.* New York: Oxford UP, 2011. Print.

Eiseley, Loren C. *The Night Country.* Lincoln: U of Nebraska P, 1997. Print.

Eliade, Mircea. *The Sacred and the Profane; the Nature of Religion.* 1st American ed. New York: Harcourt, 1959. Print.

Killingsworth, M. Jimmie. "Birdwatcher." *Interdisciplinary Studies in Literature and Environment* 16.3 (2009): 591–603. Print.

Klein, Joshua. "Joshua Klein on the Intelligence of Crows." *TED Talks.* May 2008. Web. 26 May 2012.

Leopold, Aldo, and Michael Sewell. *A Sand County Almanac: With Essays on Conservation.* New York: Oxford UP, 2001. Print.

Levy, Robert I. "The Life and Death of Ritual: Reflections on Some Ethnographic and Historical Phenomena in the Light of Roy Rappaport's Analysis of Ritual." *Ecology and the Sacred: Engaging the Anthropology of Roy A. Rappaport.* Ed. Ellen Messer and Michael Lambek. Ann Arbor: U of Michigan P, 2001. 145–169. Print.

Lincoln, Yvonna S., and Egon G. Guba. *Naturalistic Inquiry.* Beverly Hills, CA: Sage, 1985. Print.

Milton, Kay. "Ducks out of Water: Nature Conservation as Boundary Maintenance." *Natural Enemies: People-Wildlife Conflicts in Anthropological Perspective.* Ed. John Knight. New York: Routledge, 2000. 229–246. Print.

Mowat, Farley. "Floki Released a Raven." *The Bedside Book of Birds: An Avian Miscellany.* Ed. Graeme Gibson. New York: Nan A. Talese, 2005. 134–135. Print.

The New King James Bible: New Testament. Nashville: T. Nelson, 1979. Print.

Obmascik, Mark. *The Big Year: A Tale of Man, Nature, and Fowl Obsession.* New York: Free Press, 2004. Print.

O'Brien, Conan, perf. "Conan Goes Birding." *The Late Show with Conan O'Brien.* NBC, KSEE24, Fresno. 28 Jul. 2005. Television.

Peters, John Durham. *Speaking into the Air: A History of the Idea of Communication.* Chicago: U of Chicago P, 1999. Print.

Peterson, Roger Tory. *Peterson Field Guide to Birds of North America.* Peterson Field Guides. 1st ed. Boston: Houghton Mifflin Co., 2008. Print.

Price, Jennifer. *Flight Maps: Adventures with Nature in Modern America.* New York: Basic Books, 1999. Print.

Rappaport, Roy A. *Ecology, Meaning, and Religion*. Richmond, CA: North Atlantic Books, 1979. Print.

Rinfret, Sara. "Controlling Animals: Power, Foucault, and Species Management." *Society & Natural Resource* 22.6 (2009): 571–578. Print.

Rosen, Jonathan. *The Life of the Skies*. New York: Farrar, Straus and Giroux, 2008. Print.

Rothenbuhler, Eric W. *Ritual Communication: From Everyday Conversation to Mediated Ceremony*. Thousand Oaks, CA: Sage Publications, 1998. Print.

Schaffner, Spencer. *Binocular Vision: The Politics of Representation in Birdwatching Field Guides*. Amherst: University of Massachusetts Press, 2011. Print.

Spradley, James P. *Participant Observation*. New York: Holt, Rinehart and Winston, 1980. Print.

Tucker, Steven. *Bourbon, Bastards and Birds*. N.p., n.d. Web. 21 Apr. 2012.

Turner, Jack. *The Abstract Wild*. Tucson: U of Arizona P, 1996. Print.

Turner, Victor Witter. *The Anthropology of Performance*. 1st ed. New York: PAJ Publications, 1986. Print.

U.S. Department of the Interior, Fish and Wildlife Service, and U.S. Department of Commerce, U.S. Census Bureau. *2006 National Survey of Fishing, Hunting, and Wildlife-Associated Recreation*. 2006. Web. 20 May 2012.

Van Gannep, Arnold. *The Rites of Passage*. Chicago: U of Chicago P, 1960. Print.

Weidensaul, Scott. "CBC: Climate Bird Count?" *American Birds* (2009): 10–13. Print.

———. *Of a Feather: A Brief History of American Birding*. Orlando: Harcourt, 2007. Print.

Wiens, John A.,Diana Stralberg, Dennis Jongsomjit, Christine A. Howell, and Mark A. Synder. "Niches, Models and Climate Change: Assessing the Assumptions and Uncertainties." *Proceedings of the National Academy of Sciences of the United States of America* 106, Supplement 2 (2009): 19729–19736. Print.

Williams, Terry Tempest. *Refuge: An Unnatural History of Family and Place*. New York: Pantheon Books, 1991. Print.

Yoon, Carol Kaesuk. *Naming Nature: The Clash between Instinct and Science*. New York: W. W. Norton & Company, 2009. Print.

10 Banging on the Divide
Cultural Reflection and Refraction at the Zoo

Tema Milstein

Scholars who look at humanimal[1] relations often aim to raise awareness about the ways communication serves to structure, discipline and transform these relations. The core assumption in doing such work is that "how people communicate about animals helps inform the way they think about animals and shape the way they experience animals" (Milstein, "Human Communication's Effects" 1044). This chapter focuses

Figure 10.1   Girls and Gorilla. Photo by Ethan Welty.

specifically on ways communication functions to construct human relations with animals at the zoo. I look at the contemporary Western zoo to illuminate dominant discourses at animal exhibits and the subtle and not-so-subtle discursive resistances.

In particular, this study focuses on communication at the gorilla exhibit. In my research, I have found that gorillas, a major zoo visitor draw, elicit more human communication than many other exhibited animals, and this communication is filled with dialectical cultural tensions. Many of my observations are in league with Donna Haraway's argument that humans use other animals, and especially other primates, as a mirror we polish to look at ourselves and contemporary society. In zoo communication, I find various forms of struggle over meaning making that both reflect and refract clearly cultural lenses.

As intensive, public humanimal condensations, zoos as institutions strive to reflect animals in particular ways that are both culturally coherent and serve to justify zoos' continuing existence. While zoos engage in polishing particular views of exhibited gorillas—for example, that they are comfortable, playful and familial—visitors may engage in their own refractions that deflect the zoo's preferred images and construct very different views of gorillas based on connection, equivalence and emancipation. In addition, gorillas themselves engage in and influence humanimal communication. In this chapter, I focus on a particular moment when one gorilla refracts communication by the eloquent act of banging, insistently, on the zoo's human-animal glass divide.

In what follows, I explain my analytical framework, including concepts from discourse, ecolinguistics and environmental communication studies, as well as previous work on the humanature dialectical framework of mastery vs. harmony, othering vs. connection and exploitation vs. idealism (Milstein, "Somethin' Tells Me"). I situate the present study in the context of zoos as discursive sites that inform and are informed by particular cultural histories and tensions, and I describe the Western zoo site, my methodological approach and the once-celebrated naturalistic gorilla exhibit that is specific to this case study. In closely analyzing an emblematic yet exceptional humanimal communication event, I tease out reflections and refractions and discuss the ecocultural tensions and implications inherent in reaching across the symbolic-material human-animal divide.

DISCOURSE AND HUMANIMAL RELATIONS

Along with Arran Stibbe and others, I argue that human power exercised over animals and nature is materially coercive, but that this coercion is justified, reinforced, resisted and transformed in minds and institutions via discourse. Norman Fairclough, in *Discourse and Social Change*, suggests that "discursive practices are ideologically invested in so far as they incorporate significations which contribute to sustaining or restructuring power relations" (91). Discursive practices that construct animals who are subject to human control, such as zoo'd[2] animals, are thus deeply ideologically invested.

To assist in illustrating ideology within zoo communication, I use the theoretical framework of three dialectics introduced in a former study on zoo discourse (Milstein, "Somethin' Tells Me"). While the present study focuses on interpersonal communication at animal exhibits, the former focused on zoo discourse at institutional scales. Employing dialectics that emerged from an institutional-scale examination to explore interpersonal scales of communication may add a depth of dialogic understanding to the multifaceted and interlocking ways humanimal relations are constructed in zoo settings.

Examinations of humanature discourses at the interpersonal and even intrapersonal scale often reveal multiple ideologies in dialogue (Marafiote and Plec). As dominant discourses assert themselves, counterdiscourses often interweave within the dominant ones, at times challenging dominant perceptions and practices. Similarly, I argue that the three tensions within zoo institutional discourse, the dialectics of *mastery-harmony, othering-connection* and *exploitation-idealism*, may be found at the interpersonal and intrapersonal scale. I briefly define how these dialectics serve as a context for analysis.

The first dialectic's dominant pole is *mastery*, an ideology that takes human control over nature and other animals as both a given and a precondition of societal progress. Mastery's counterdiscourse—*harmony*—values holistic cooperation and positions so-called progress, such as industrialization, corporatization and neoliberal globalization, as damaging ecological balance and any possibility of harmony. The second dialectic's dominant pole is *othering*, differentiating humans from other animals, nature and at times Othered humans. In this fundamental dualism of Western cultures (Carbaugh; Plumwood, "Androcentrism and Anthropocentricism"), the center (e.g., humans, whites, socioeconomic elites, men, heterosexuals) subordinates the other (e.g., animals, minorities, poor people, women, nonheterosexuals), justifies oppressive views and practices and obfuscates knowledge that humans are, in fact, animal and natural. The counterdiscourse is *connection*, which refracts dualisms and positions humans as interdependent forces interacting with other ecological, sensual, emotional and comprehending forces. The final dialectic's dominant pole is *exploitation*, in which nature's value is in its instrumental commodification for human gain or pleasure. The counterdiscourse is *idealism*, which circulates desires to create alternative realities by preserving, restoring and respecting humanature for its intrinsic value. In its damaged form, blindered idealism can allow for rationalizations of exploitation, but in its creative ecocentric form idealism can override the logics of domination that "create 'blind spots' in the dominant culture's understanding of its relationship to the biosphere . . . " (Plumwood *Feminism* 194).

These three dialectics—mastery vs. harmony, othering vs. connection and exploitation vs. idealism—centrally inform Western discourse, yet the dialectical poles do not receive equal public discursive consideration. Dominant profit-driven Western cultural practices, such as excessive mass consumption, largely rest upon capitalization of nature, nonhuman animals and marginalized people via material-symbolic practices of mastery, othering and exploitation. Thus, even in cultural settings such as zoos (which, in their most recent iterations, are ostensibly dedicated to the sustainability of animals, humanature and the planet) the counterthemes of harmony, connection and idealism tend to be foregrounded yet ultimately subordinated themes.

To aid in my analysis, I also look to extant concepts such as James C. Scott's hidden transcript of the oppressed, which, spoken openly, disrupts dominant discourses and power relations; Val Plumwood's notion that humans at times are able to speak strategically in liberating ways for nature or for animals stripped of their voices ("Androcentrism and Anthropocentricism"); and Rom Harré, Jens Brockmeier and Peter Mühlhäuser's argument that Western syntax largely positions humanature relations as causal, with human as agent and animal and nature as objects, discursively obliterating nonhuman nature's agency and subjectivity.

ZOOS AS SITES OF MATERIAL-SYMBOLIC DISCOURSE

A dialogic dialectical lens suits zoo studies, as zoos are in-between places of tension in Western ecoculture. As public institutions, they occupy the liminal spaces between recreation and education, science and showmanship, high and low culture, remote nature and cityscape and wild animals and urban people (Hanson). David Hancocks argues that zoos present dichotomies of confused, cold, captive conditions vs. sensorial, emotional and even rehabilitating places of wonder. As such, zoos "reveal the best and the worst in us and are stark portrayals of our confused relationship with the other animals with which we share this planet" (xvii).

Zoos are also widely popular, annually drawing 150 million visitors in the United States, more than the combined U.S. annual attendance of professional football, basketball, baseball and hockey (aza.org). In order to maintain popularity, zoos have had to transform with the culture and critiques of their times. In the past forty years, zoos, to some degree, have reinvented themselves, particularly in the areas of animal exhibition, treatment and conservation. In the process, many leading zoos have recast themselves as quasi-natural, replacing viewing bars and concrete with glass-window divides and fabricated slices of simulated habitats.

These changes are as much, or more, for the human visitor as for the exhibited animal.[3] Visitors of naturalistic exhibits look through portholes into wild-like virtual habitats (Bostock). These views are intended to ameliorate any dissonance over viewing animals in captivity and to stimulate respect and admiration rather than pity, superiority or displays of mocking cruelty (Hancocks). As such, contemporary zoos often actively endeavor to avoid positioning the zoo animal as spectacle, instead displaying animals in nature-like environments in which they appear to have privacy, autonomy and the ability to avoid the human gaze (Davis).

Yet the zoo is not merely shaped by the discourse of its time; the institution is also a discourse in itself, shaping the humanimal relations within. Beardsworth and Bryman outline the genealogy of today's zoos, illustrating how in historical and contemporary forms, the zoo has always carried two fundamental themes: gaze and power. In exhibiting animals, zoos' central function is a process of power wherein "almost total control is exercised by humans over animals' movements and activities, with minimal opportunity for the animal to exercise its own preferences or priorities" (88). As such, though some zoo visitors may intend and attempt subject-to-subject encounters with zoo animals, Kaplan argues the viewing gaze itself is inextricably linked to objectification, making the exhibited animal the object of constant scrutiny.

The naturalistic exhibits in themselves serve to disguise the inconsistency of dialectical impulses. Baratay and Hardouin-Fugier argue that the design of the contemporary zoo expresses the illusion of certain natural spaces and represents wished-for environmental ideals while the global plunder of natural habitats intensifies unabated. Wilson argues that, in zoos' transformations from spaces largely reflecting displacement and domination to spaces ostensibly more concerned with simulated habitat and environmental education, zoo concerns have shifted from a focus on the method of containment of zoo'd animals to an examination of human viewing of zoo'd animals, and of relationships among nonhuman species. This shift parallels ecocultural shifts in humanature and humanimal spheres outside zoo walls. Indeed, today's zoos "no longer represent the vastness of empire or the abundance of the natural world. Today the inhabitants of zoos are often the last remnants of a species or community. Their exoticism is an exoticism of imminent loss" (Wilson 247).

Yet power is still foundational in the process of exhibition, even at the most basic level: the human being is always free to leave and the zoo'd animal is always confined (Jamieson). The power of exhibition serves to favor and protect the human visitor in less obvious ways, as well. The zoo visitor views zoo'd animals, gains pleasure, knowledge and/or entertainment, regardless of the animals' preferences or desires. At the same time, due to the animals' captive state, the visitor generally remains entirely protected from feelings and realities of reciprocity or vulnerability in the humanimal intersection. Along these lines, Berger asserts zoo'd animals essentially disappear, incapable in the subordinated setting of equally reciprocating the visitor's gaze. While zoo'd animals' capability to reciprocate the gaze may be argued, elsewhere I contend it is apparent that in "their surveilled captivity, the vast majority of zoo animals have been immunized from engaging in actual encounters with visitors" (Milstein, "Somethin' Tells Me" 33).

ZOO WEST'S NATURALISTIC GORILLA EXHIBIT

The material-symbolic contexts of zoos outlined above and the dearth of subject-to-subject humanimal encounters at zoos make the communication event I analyze here intriguing as a point of analysis. The event, a zoo staff-guided elementary school tour of a gorilla exhibit at Zoo West,[4] both substantiates and complicates claims about zoos and animal-visitor encounters. This tour was one of many communication events I observed during two years of participant observation on the human side of the gorilla exhibit's glass divide. I chose this particular text as analytical focal point because it is both exceptional and exemplary of core themes and discursive struggles I observed in my fieldwork. What made this event unique, and in my opinion worthy of analysis, were the actions of one young gorilla, actions that initiated subject-to-subject humanimal encounters and a tension-rich dialogue among tour guide, children and gorilla.

My approach is informed by Donal Carbaugh's assertion that the case study enables comparative assessment of available human discursive means for understanding and evaluating nature as well as an analysis of the attendant attitudes such discourses may cultivate or constrain. In order to better analyze these cultivations and constraints, I use an analytical framework of *reflections* and *refractions*, two cultural counterparts rooted in the dialectical structure used here as theoretical framework. I apply these analytical terms to humanimal communication at the gorilla exhibit in order to draw out powerful pulls between dominant and counterdiscourses. In doing so, I attempt to illustrate the ecocultural tensions at play in interpersonal communication at the zoo.

In my interpretations, I found the tools of critical discourse analysis apt for examining the powerful ways humans symbolically cage or free other humans, animals and nature. Critical discourse analysis, like other critical methodologies, identifies its object of study within a web of interrelationships and power (Fairclough, *Analysing Discourse*). In looking at structural or situational concerns, analysts look at the social event in which the text is observed (in this case, a gorilla's initiation and complication of communication during a schoolchildren's tour at the exhibit), the genre in which the text is situated (the guided zoo tour), and the discourses, styles and identities drawn upon in the text, which I illustrate below in my analysis. I use tools of critical discourse analysis, such as a close look at lexicon, syntax and particular logics produced in discourse, to analyze my data.

The setting for this case study is a major urban zoo in the U.S. West, often viewed as an early pioneer of naturalistic exhibits that set standards for zoos around the world. Before such changes began in the mid-1970s, Zoo West had a traditional exhibit approach, grouping animals by species and exhibiting without ecological context to provide the best close-up views. In the 1970s, along with other leading zoos, Zoo West began immersing visitors in simulated landscapes with living and artificial flora that attempted to approximate habitats. Zoo designers demolished some older moat cages, dropped visitors to equal or lower viewing levels, and replaced bars with glass or no immediately obvious barrier. One intention behind these changes was to reduce human-zoo'd animal power differentials.

In 1978, in a celebrated and, early on, controversial move, Zoo West's endangered lowland gorillas became the first zoo-confined gorillas in the world to move from small concrete cages to a naturalistic "habitat." Dian Fossey and other gorilla researchers assisted in the design of this first gorilla "landscape-immersion" exhibit, and for some time Fossey told audiences that Zoo West had the only zoo facilities suitable for keeping gorillas (Hancocks). Such exhibits, now globally popular, were intended to immerse visitors in a realistic illusion of wild gorilla worlds.

Zoo staff noticed differences in both gorillas and visitors after the exhibit change. The aggressive behavior of gorillas toward one another decreased and extreme inertia gave way to relaxation, exploration, play and apparent contentment. On the other side of the glass divide, human visitors

> who once had stood in the grimy corridor of the old ape house, passively gawking or mocking the animals with whoops and shuffling jumps, now stood in small clearings amid dense vegetation and did not shout or howl or, often, even talk, but occasionally whispered to each other, with wonder in their eyes. (Hancocks 134)[5]

Wilson describes some details of the groundbreaking Zoo West gorilla exhibit: a simulated tropical forest clearing that demonstrates the process of plant succession; a glass-window lean-to at one side that is the vantage point for visitors; a series of boulders heated by electric cables that in the cold of winter often make it necessary for the gorillas to sit directly in front of the window; more distant boulders and a stream and group of caves to which gorillas may escape—these relate to flight distance, the nonsocial space animals need to feel safe. Wilson states these developments were encouraging, but argues that questions remain: "Do the new designs somehow disguise the confinement that is the primary fact of the zoo? Do wild-animal displays conceal and mystify the ways some human cultures continue to dominate the natural world? Can we really see ourselves looking?" (254).

Similar questions guide this study, as does recent research on zoo-visitor knowledge and attitudes toward gorillas and chimpanzees. For instance, Kristen Lukas and Stephen Ross found no difference in visitor attitudes after zoo visits. As such, attitudes toward gorillas that prior to the visit were more negative than attitudes toward chimpanzees remained negative. The authors argue there is room for improvement in how zoos, via their exhibiting of great apes, engender a conservation ethic and motivate conservation action.

Forty years after Zoo West's gorilla exhibit became the model for zoos, Wilson's questions and the issue of whether zoo exhibits actually contribute to sustainable human-gorilla perceptions and relations are still pertinent and help guide my interpretations of contemporary communication at the gorilla exhibit. Today, the silent or whispered contemplations Hancocks mentions are no longer the norm. Zoo visitors are acclimated to naturalistic exhibit design and largely take it as a given. A range of communication can now be observed at naturalistic exhibits, from mocking to respectful. Yet the immersive exhibits also appear to elicit more complex, nuanced and powerfully negotiated humanimal discourse.

THE TOUR

As mentioned, I analyze a particular instance of dialogic communication; in this event a young gorilla, along with child participants, initiates, interrupts and transforms the dominant zoo discourse provided by the tour guide. The naturalistic exhibit design can be seen as playing a large role in this exchange, plunging the tour into gorillas' simulated habitat and, via the glass divide, providing gorilla, children and tour guide at once both intimate access to—and separation from—each other. I provide the complete text from a transcribed tape recording of

Figure 10.2 Akenji Pounds on the Glass. Photo by Tema Milstein.

the class tour during its stop at the exhibit, followed by my analysis.[6] The tour guide (G), as employee of the zoo, director of the tour and lead adult among a group of mostly children, has extensive power as the social agent who textures the discourse. A few adults, both teachers and parent volunteers, are present with the children but do not speak. When the class tour arrives, Akenji, a two-and-a-half-year-old gorilla, does something unprecedented in my study observations. Whereas the gorillas generally ignored human visitors, casting a seemingly disinterested glance now and then, Akenji runs to the glass divide between her and the children and begins pounding on it. She does not stop until after the children have departed.

1　Children: ((bubbly laughter))
2　G: And we have another baby Akenji visiting with her mama and they may
3　knock on the glass that's ok it's *their* glass they can knock on it if they want but we
4　would never bang back right yes this baby's being really cute over here let's keep our
5　voices down
6　Children: ((bubbly laughter builds as Akenji stands facing them two feet away, hands
7　high, banging on glass))
8　G: The male is up and kind of moving around now sitting looking right at us with her
9　little cane in front of her there ((laugh)) with her tongue kind of sticking out a little bit
10　that is Nina and she is the grandmother so she's the mother of this mom and baby
11　lying right here on her back alright and Akenji is going to entertain us here or excuse
12　me Naku is going to entertain us here
13　Child: Who's that?
14　G: =And the one sitting down close to the glass over here that's Nadiri we call her
15　Nadi sometimes and she's about seven years old
16　Child: I'm eight
17　G: *Yeah* I think she's about ready to turn eight or maybe she just did ok now you get to
18　see the other baby for a second aww she's going to curl back up with mom (.) and
19　Akenji's over here just having herself a good old time yeah Nina just looks like a
20　grandma doesn't she
21　Child: yeah

22 G: Does she look like a grandma or what it's like she looks up she heard me say that

23 ((laugh))

24 Child: <question about heat lamps>

25 G: Yes yes now these gorillas live in Africa and it's pretty hot there right

26 Children: yeah

27 G: Yeah do you think they would be very comfortable just out in the cold in our

28 climate in the winter

29 Children: no

30 G: So look up above they all have heating lamps and that's one of the reasons they

31 like to sit in this nice warm *dry* area and of course the keepers clean it for them

32 everyday it's *not* uh this baby over here's just playing up a storm on the window

33 Children: ((laughs))

34 G: If we gave her a drum set it might be really interesting to see what she'd do alright

35 let's see what dad's doing

36 Girl watching Akenji: Maybe he wants to be let out

37 G: Yeah you *think* so I think she's just playin'

38 Another child: So, she's trying to get the lock undone

39 Other children: eeeew ((laughter))

40 G: I think she's just showing off for you would you guys want to leave it's a *beautiful*

41 environment they get fed everyday=*hey* let's talk about what they eat now look around and

42 you can see in their exhibit you see some branches with some leaves on them and

43 they like to eat those leaves but [ ]

44 Child: *Hi* ((to Akenji))

45 G: We also give them celery and I think they get some carrots sometimes and

46 maybe some fruit now and again and occasionally they like a little bamboo

47 Akenji: ((continuing to pound on glass))

48 Children: ((bubbly laughing))

49 G: And she's just having a good ol' time here pounding away alright ALRIGHT my

50 eagles we're going to move along EAGLES this way alright we're going to go back

51 past the jaguar exhibit so you get one more look at him

52 Akenji: ((still pounding))

REFLECTIONS

I reserve my analysis of the discourse surrounding Akenji for the following section and focus in this section on dominant zoo discourse as constructed by the guide. Subject positioning in this text is persistent as the guide, with the authority deriving from her position as representative for the institution and as lead adult, has the dominant power to shape communication about the gorillas. The guide's talk is informally scripted, loose enough to shift a bit when surprises, such as Akenji's banging on the window, crop up. The guide consistently strengthens her authoritative voice by including the appearance of dialogism, introducing closed-ended questions that have automatic, obvious yes-or-no answers, which she either answers herself (e.g., lines 3 and 4 "they can knock on it if they want but we would never bang back right yes") or which children answer automatically with pat answers (e.g., lines 19 and 20 "Nina just looks like a grandma doesn't she" Child: "yeah"; lines 25–29 G: "now these gorillas live in Africa and it's pretty hot there right" Children: "yeah" G: "Yeah do you think they would be very comfortable just out in the cold in our climate in the winter" Children: "no"). This style incorporates a traditional pedagogical genre, in which lessons are given to students in a way that appears as if student voices are included, but in fact the children are mouthing forecasted answers.

The guide draws on the othering-connection dialectic, using pronominalization to divide gorilla from human with the exclusive "we" for humans and "they" for gorillas, differentiating gorillas (and what they may do with "their" glass divide) from humans (and what we may not do with "their" glass divide). Other pronominalizations occur in lines 27–31, in which differences are established between visitor and gorilla climates (e.g., "do you think they would be very comfortable just out in the cold in our climate"), followed by a statement of fact about what "they," the gorillas, like. The guide uses the logic of appearances to say gorillas "like to sit in this nice warm *dry* area." As such, she anticipates potential visitor connective concerns for gorilla welfare in a cold, wet climate. She eschews explanatory logic, which would represent gorillas as being far removed from their African tropical rainforest home, placed in this foreign climate and, therefore, forced to sit under heat lamps located only in front of the glass divide. If they are to keep warm, gorillas are constantly on display for the visitors' gaze. Here, we see the emergence of the exploitation-idealism dialectic. The guide's use of the logic of appearances falsely represents an ideal situation of gorillas sitting under lamps as a "like" instead of a "need," masking exploitive design elements of the exhibit and implying that gorillas want to be near humans—a conflicting and potentially dangerous message in terms of the zoo's overarching conservation packaging.

Lexicalization, or choice of words, also serves to conceal exploitive, mastery and othering discourses. Zoo discourse selectively represents human agency only in idealistic, harmonious and connective acts of looking out for gorilla welfare. For instance, "we," or the human center, "give them" food, but "we" do not put gorillas in this captive situation to begin with through destruction of gorilla habitat, individual violence done to them by poachers or catching and caging. Instead, humans, as presented in the tour, figure as agents only as keepers and stewards making sure the gorillas' area is clean, warm and beautiful, and that they are fed.

Harré, Brockmeier and Mühlhäuser argue that Western syntax largely positions humanature relations as binary and causal, erasing nonhuman nature's subjectivity and agency. In the case of the zoo, passivization reflects a material day-to-day existence for captive animals cut off from agency. In lines 40–46, however, the guide mystifies this passivization via syntactical arrangement. Whereas foraging for food, as gorillas would in their ecological habitats, is impossible in exhibits, the guide strategically positions gorillas as agents with desires met by the zoo. Statements such as "[they] "like to eat those leaves" and "occasionally they like a little bamboo" signify a choice of diet these gorillas lack. Similarly, a sign, posted next to the glass of the exhibit, depicting two gorillas foraging in the wild and the singular term "gorilla," furthers this mystification, as do live plants inside the exhibit that do not serve as a food source.

The mastery-harmony dialectic again emerges in representations of the gorillas' social world. Gorillas are represented as members of a harmonious extended family which closely resembles a human family, with grandmothers, moms, dads and babies. The gorillas are one animal group in the zoo who are, in fact, mostly blood related. The oldest gorillas were wild and captured as babies and, now middle-aged, have long found themselves used as breeders in a planned nationwide zoo breeding program. However, this mastery-informed, human-controlled gorilla procreation is obscured. Unmentioned to the children are the absent offspring permanently removed to other zoos for breeding.[7] Harmony is also favored in avoidance of mention of birth control, though this exclusion is not surprising considering the child audience. Birth control does come up at times when exhibit docents speak to adult visitors, yet these references are generally too vague for most adults to understand that female gorillas are on birth control largely to keep from getting pregnant by their fathers in this unnatural population situation of the zoo. In addition, the moving of one baby gorilla from another gorilla exhibit because of adult aggression and violence is not generally discussed with visitors.[8]

As mentioned, Haraway argues that nonhuman primates serve as a mirror humans polish to assemble images of themselves and human society. A representation of a gorilla family that includes incest, violence, controlled breeding and offspring taken away from parents would provide a very different image, and hence reflect quite negatively on humans. The tour guide thus makes ideological lexical choices in highlighting relations in terms of harmonious familial relations (e.g., line 35 "let's see what dad's doing"), and excluding other poignant details.

Naming is also striking in this text. The gorillas are named with African human names, such as Akenji and Nadiri. The exception, ironically, is Nina, the only gorilla mentioned by name in the tour who was actually born in Africa. Nina was captured well before zoo naming practices favored names of humans who live in proximity to gorillas' wild habitats. Cultural differences are represented via these names, which is in line with the zoo's stated intention to help visitors see connections between gorillas and their natural habitat, and to encourage visitors to feel like protecting the animals' complete ecosystem of interrelations. However, these naming practices can also mask exploitation, mastery and othering with the idealism, harmony and connection dialectical poles, camouflaging born-in-Western-captivity gorillas by associating them with culturally marked people.

In contrast to Haraway, Morbello argues that humans are not invited to see themselves reflected in the zoo-animal mirror. Instead, visitors who look in the culturally themed zoo mirror see reflected back those people marked and exoticized as other. The design of naturalistic animal exhibits serves to further conceal obvious signs of Western zoo artifice or impact, such as keeper doors or locks. Such material-symbolic discourses leave the zoo'd animal othered and exotic, and leave the visitor immersed in untouched illusionary nature and with blinders to overseas gorilla habitat destruction.

Whereas the overarching genre of the text examined here is that of a guided tour, intertextuality and recontextualization are also at play. The guide switches genres, mixing a pedagogical genre and a kind of nature show genre à la *Wild Kingdom*; one can almost hear a hushed narrative being delivered as gorillas wander through the forest, such as in line 8: "The male is up and kind of moving around." In the next sentence, the guide switches to a more informal anthropomorphizing genre pointing out the "cane" of Nina, the "grandmother": "now sitting looking right at us with her little cane in front of her there ((laugh)) with her tongue kind of sticking out a little bit that is Nina." The hybridization of these genres helps legitimize the guide's anthropomorphizing statements, as most Westerners are accustomed to turning to zoo guides, teachers and nature documentary shows for information about nonhuman animals in their natural habitats, and these genres often incorporate anthropomorphism. The use of the nature show genre also mystifies the captive state of animals in a fabricated exhibit, shifting the visitor's gaze to look through the exhibit glass as though it were a giant television screen giving a glimpse into a wild habitat. I now turn to the reframing of this glass divide.

REFRACTIONS

Perhaps the most striking aspect of this text is the discourse surrounding the pounding of the glass by Akenji. The young gorilla draws a different kind of attention to the transparent divide between human visitor and exhibited animal and, throughout the tour, relentlessly returns attention to this divide. Akenji's actions appear to jar the guide, who works equally relentlessly to represent Akenji's behavior in dominant zoo discourse largely steeped in mastery, othering and exploitation. At the same time, Akenji's actions appear to connect with the children, who discursively represent Akenji's actions quite differently, in communication steeped in harmony, connection and idealism. The dialectical competition between interpretations offered by the guide and by the children is complicated by their relative power within the discursive regime of the zoo.

The guide first represents Akenji as "visiting" the children's tour. This first humanimal connection framing, however, is then replaced by the guide's persistent discursive work to represent Akenji's actions as playful, fun and performative. In these repeated representations, the guide separates and distances Akenji as entertainer for the human audience: "this baby's being really cute over here" (line 4), "Akenji is going to entertain us here . . . " (lines 11 and 12), "this baby over here's just playing up a storm on the window" (line 32), "If we gave her a drum set it might be really interesting to see what she'd do" (line 34).

With all these dominant representations of Akenji's actions, it would seem that the guide anticipates the refractions just below the surface. The first refraction is provided by a child in line 36: "Maybe he wants to be let out." It is notable that this empathetic statement is in direct opposition to the guide's framing, as if the child heard too many of the guide's representations and finally burst out with an alternative meaning. The child's translation of Akenji's communication as possibly representing a desire for freedom implies connection to another being's desires and needs within a restricted world. The guide then does quick work in reclaiming the authority to represent, and to deny the child's authority, using her subject positioning to overpower the child's interpretation (line 38: "Yeah you *think* so I think she's just playin'"). The guide's positioning compared to that of the child and her verbal emphasis on the child's "think" and deemphasis on her own "think" differentiate the weight and accuracy of each of their statements, subordinating the girl's "think" (Akenji wants to be let out) to her own (Akenji is just playing).

As the children continue ascribing an alternative connection-based meaning to Akenji's actions, the guide uses her positioning and louder adult voice to continue the work of framing, using a damaged form of idealism and her superordinated "think" to do the work: "I think she's just showing off for you would you guys want to leave it's a *beautiful* environment they get fed everyday" (lines 40 and 41). And then quickly, without a pause, the guide changes the subject, attempting to redirect attention: "=*hey* let's talk about what they eat now look around . . . " (line 41). The guide's representation of Akenji's behavior as "showing off" for the children exhibits a mastery orientation of zoo'd animals' purpose, an exuberance for entertaining humans. At the same time, the guide attempts to shift the children's gaze from Akenji to the less problematic docile adult gorillas and to shift the children's thoughts to what gorillas "like." Notably this interlude is the one time the guide asks a question—"would you guys want to leave"—for which she is not assured of the children's answer. The guide, without pause, then changes the subject, closing the space that might allow opportunity for a dissenting, or refracting, answer.

Interestingly and oppositionally, it is at this point that one of the children facing Akenji speaks from a harmonious stance as the guide speaks, directly addressing Akenji, to whom no one has thus far spoken: "*Hi*" (line 44). Both children introduce resistant discourses, re-presenting Akenji's actions and appropriate humanimal interaction. Each favors the harmony, connection and idealism poles of the dialectics in place of the mastery, othering and exploitation poles that better suit the psychic and capitalist needs of the zoo. The first child's refraction evokes recognition of captivity and desires for freedom. The second child's evokes equivalence and harmony with another animal, a respectful return or initiation of greeting.

Earlier in the transcript, another statement of equivalence is found in line 16, when a child responds to the guide's naming and identification of the age of a gorilla who is the child's age by stating, "I'm eight." This after the guide states differences as to who may touch the glass divide: "it's *their* glass they can knock on it if they want but we would never bang back right yes" (lines 3–5). While it is clear that the guide would say this in order to protect the gorillas, at the same time this statement preemptively others the gorillas while perversely also attributing ownership of the glass to them. While the children immediately and continuously react to Akenji's banging with boisterous and connective excitement, they are admonished to "never bang back." By not providing a way to respond (such as waving) and not acknowledging the children who make statements of connection, the chasm widens as the children are taught not to respond to the expressive outpouring of another animal.

Despite the children's reframing, the guide has the final human word (Akenji has the final zoo'd animal word, continuing to pound on the glass), as well as the physical mastery and control over the children in deciding when they are to stay in this humanimal space and when they are to leave. The guide's final reflection points one last time to the fun Akenji is having, legitimizing the guide's dominant representation and then removing the children from the sight and sound of Akenji: "And she's just having a good ol' time here pounding away alright ALRIGHT my eagles we're going to move along EAGLES this way . . . " (lines 49–50).

MAKING VISIBLE, UNSETTLING AND
REACHING ACROSS THE DIVIDE

In reading the transcript, the first question might be: Why does the guide appear to be discursively working so hard? I argue that in banging on the divide, Akenji sends tremors through the polished junctures of Western human-animal and culture-nature binaries. In simultaneously drawing attention to the divide and reaching across it with her young human inter-locutors, Akenji reveals a material-symbolic gulf the zoo works to natu-ralize and maintain. In the face of Akenji's disruption and the children's responsive refractions, the guide struggles to reestablish discursive footing. In doing so, she navigates dialectical tensions, attempting to reassert zoo reflections of happy animals on exhibit over the unsettling refractions of children and gorilla.

Yet Akenji and the children appear to succeed in making the divide visible. And, by communicating through the divide to one another, they take the next step that could follow an awareness-raising action like Aken-ji's—they attempt and, one could argue at least discursively, effect change. Instead of emphasizing the mastery-othering-exploitation poles of the dia-lectics (as does the guide), in emphasizing equivalence, empathy and reach-ing across the divide, Akenji and the children favor and put forth harmony, connection and idealism.

As mentioned, according to zoo literature, glassed exhibits serve multiple purposes. Glass replaced bars in most contemporary zoos in part to deem-phasize captivity for the visitor and provide the effect of a porthole into a habitat. However, the introduction of glass in zoos also further bodily and sensually separates human zoo visitor and animal detainee. In 1961, when a different era of cultural discourses were informing leading zoo deci-sions, early glass fronts added to interior cages were championed as serving to restrict "offensive odors" to the zoo'd animal space and human-carried disease to the human (www.zoo.org). The divide served both symbolic and material intentions, at once connecting and dividing. This is especially poi-gnant when one considers the important role scent plays in much animal communication. It may be a kind of sensory deprivation torment to be largely cut off from the information provided by one's senses of smell, taste, touch and sound, and limited to one's sight to perceive the world on the other side of the glass divide.

Today, drawing attention to the divide, either symbolically or materially, is not in the interest of the zoo's livelihood, in which captivity is the elephant in the room, so to speak. Children and gorillas, however, are not culturally invested in zoo or wider humanimal reflections. Children are far more likely to perceive veracity and to voice injustice than are economically invested institutions or culturally invested adults. Indeed, children, in their innocence, often point to the obvious fact clothed in the cultural conceit, ranging from the naked emperor on parade to the captive gorilla communicating.

James Scott calls such disruptions of the dominant discourse open state-ments of a hidden transcript. Scott argues that the first open statement of a hidden transcript breaches the etiquette of power relations, breaks open an apparently calm surface of silence and consent, and carries the force of a symbolic declaration of war. Akenji's and the children's communication comprise a hidden transcript in the sense that their refractory messages remain unstated by the vast majority of adults who visit the zoo; yet such messages are on many adult minds. In contrast to Scott's notion of the hidden transcript, which is spoken by subordinated humans about their own lot, when it comes to subordinated nonhuman animals, it is often humans who must speak if other humans are to hear.

In this case, as the guide repeatedly reasserts the dominant reflection, Akenji upsets the apparently calm surface and children voice the hidden transcript. In another study on endangered whale watching, I've argued that such strategic speaking for nonhuman animals is not limited to children (Milstein, "Nature Identification"). Adults, too, may use communication strategically to speak for animals culturally stripped of their voice, allowing for more sustainable or even restorative humanature refractions. In this case, however, the guide (and perhaps the other present but silent adults) is directly invested in the same dominant discourses in which the zoo and many other Western institutions are invested, discourses that largely pivot upon mastery, othering and exploitation.

While Akenji and the children bang on the divide, the guide mystifies ownership of the glass. The glass divide is not a human-emplaced barrier, but rather the gorillas' "glass they can knock on it if they want but we would never bang back right yes." The obfuscation of mastery, othering, exploitation and the passivized empowerment of gorillas is furthered by the guide's strategic syntactical positioning of zoo'd gorillas as agents, as opposed to unconsenting captives and victims of unsustainable global human practices. A slave-master narrative is evidenced throughout the guide's discursive reframing: Why would they want to leave? We're nice to them, give them a nice space, feed them, etc.

Via the zoo representative's mouth, gorillas remain unvoiced agents— with the safe exceptions being grandmothers with canes or young entertainers who would be better off with drum sets. Indeed, the guide, in her authoritative zoo voice, always has an answer to affix a particular reflection. Entertainment, not emancipation; separation, not equivalence.

I've chosen to build upon Haraway's notion of the mirror to elucidate a humanimal interplay among discourse, reflection and refraction. The material-symbolic presence of the zoo-exhibit glass serves at once as window, divide and mirror, a surface we discursively polish to shape the animals on both sides. While reflections such as the guide's can be considered the throwing back of a sought-after constructed state of being, Akenji's and the children's communicative refractions can be seen as deflections. The children and Akenji turn or bend dominant constructs much as refracted waves of light or sound can turn or bend when passing through one medium to another. Such changes in direction are potentially stronger than acts of outright resistance, as they do not simply push back against a force—a resistive approach that can often lead to standstill or defeat—but instead take existing shared energy and move it in a different direction.

Figure 10.3 Sleeping Gorilla. Photo by Ethan Welty.

NOTES

1. I use the compound terms "humanimal," "humanature" and "ecoculture" throughout my writing as a way to reflexively engage human and animal, human and nature, ecology and culture, in integral conversation in research as they are in life. These discursive moves are turns away from binary constructs and notions of humans as separate from animals, nature and "the environment" and turns toward a lexical reciprocal intertwining reflective of living symbolic-material relations. Milstein, Tema, Anguiano, Claudia, Sandoval, Jennifer, Chen, Yea-Wen, & Dickinson, Elizabeth. "Communicating a "new" environmental vernacular: A sense of relations-in-place." *Communication Monographs*. 78 (2011): 486–510. Print. The terms are in league with Haraway's use of "naturecultures" to encompass nature and culture as interrelated historical and contemporary entities (*When Species Meet*).
2. Changing the adjective in "zoo animals" to the verb in "zoo'd animals" is my attempt to do the discursive work of pointing to an active process in which humans are the implicit agent. Verbing passive adjectives that naturalize particular humanature relations (such as "zoo," "farm," "pet" and "laboratory") is one step toward making visible particular ecocultural relations as active constructions with material consequences and possible alternatives.
3. Simulated naturalistic changes to exhibits may have some impact on the animal's mental, biological and emotional health, yet zoo professionals point out that most zoo animals spend much of the day and all night in barren concrete and steel cages. David Hancocks states that conditions publicly criticized in the past are still the norm, but are now merely out of sight.
4. Pseudonym.

5. As such, Hancocks argues that zoo-exhibit design subconsciously affects visitor perception of animals. Yet very few studies empirically examine perceptual or behavioral changes linked to differences in exhibit. One exception is a study at Australia's Melbourne Zoo that found visitors viewing gorillas in a concrete pit cage in 1988 chose predominantly negative words to describe them, such as "vicious," "ugly," "boring" and "stupid," whereas, two years later after relocation to a large African rain forest–simulating naturalistic exhibit, visitors chose very different descriptors–such as "fascinating," "peaceful," "fantastic" and "powerful." Amanda Embury, "Gorilla Rain Forest at Melbourne Zoo," *International Zoo Yearbook*, vol. 31 (London: Zoological Society of London, 1992).

6. This particular tour took place on March 3, 2004, mid-day at the gorilla exhibit. There are two holding areas of gorillas at the zoo. I was at the west exhibit. Underlined words indicate emphasis on the part of the speaker.

7. With gorillas, the zoo usually moves male offspring as one adult male gorilla per group is the maximum possible in exhibit conditions.

8. Some docents shared this kind of information with me while I observed at the gorilla exhibit as a researcher (I wore a researcher name tag), likely because I expressed curiosity and often spoke to them at some length. As far as I was able to observe, however, zoo staff and volunteers rarely spoke of such things to regular zoo visitors.

REFERENCES

Baratay, Eric, and Elisabeth Hardouin-Fugier. *Zoo: A History of Zoological Gardens in the West*. London: Reaktion Books, 2002. Print.

Beardsworth, Alan, and Alan Bryman. "The Wild Animal in Late Modernity: The Case of the Disneyization of Zoos." *Tourist Studies* 1.1 (2001): 83–104. Print.

Berger, John. *About Looking*. New York: Random House, 1980/1991. Print.

Bostock, Stephen St. C. *Zoos and Animal Rights: The Ethics of Keeping Animals*. London: Routledge, 1993. Print.

Carbaugh, Donal. "Naturalizing Communication and Culture." *The Symbolic Earth: Discourse and Our Creation of the Environment*. Ed. James G. Cantrill and Christine L. Oravec. Lexington: UP of Kentucky, 1996. 38–57. Print.

Davis, Susan. "'Touch the Magic.'" *Uncommon Ground: Rethinking the Human Place in Nature*. Ed. William Cronon. New York: W. W. Norton & Co, 1996. 204–217. Print.

Embury, Amanda. "Gorilla Rain Forest at Melbourne Zoo." *International Zoo Yearbook*. Vol. 31. London: Zoological Society of London, 1992. Print.

Fairclough, Norman. *Analysing Discourse: Textual Analysis for Social Research*. London: Routledge, 2003. Print.

———. *Discourse and Social Change*. Malden, MA: Blackwell, 2002. Print.

Hancocks, David. *A Different Nature: The Paradoxical World of Zoos and Their Uncertain Future*. Berkeley, Los Angeles, London: U of California P, 2001. Print.

Hanson, Elizabeth. *Animal Attractions: Nature on Display in American Zoos*. Princeton and Oxford: Princeton UP, 2002. Print.

Haraway, Donna. *Simians, Cyborgs, and Women: The Reinvention of Nature*. London: Free Association Books, 1991. Print.

———. *When Species Meet*. Minneapolis: U of Minnesota P, 2008. Print.

Harré, Rom, Jens Brockmeier, and Peter Mühlhäuser. *Greenspeak: A Study of Environmental Discourse*. Thousand Oaks, CA: Sage P, 1999. Print.

Jamieson, Dale. "Zoos Revisited." *Ethics on the Ark: Zoos, Animal Welfare, and Wildlife Conservation.* Ed. Bryan G. Norton et al. Washington, D.C., and London: Smithsonian Institution P, 1995. Print.

Kaplan, E. A. *Looking for the Other: Feminism, Film, and the Imperial Gaze.* New York: Routledge, 1997. Print.

Lukas, Kristen E., and Stephen R. Ross. "Zoo Visitor Knowledge and Attitudes toward Gorillas and Chimpanzees." *The Journal of Environmental Education* 36.4 (2005): 33–48. Print.

Marafiote, Tracy, and Emily Plec. "From Dualisms to Dialogism: Hybridity in Discourse about the Natural World." *The Environmental Communication Yearbook* 3 (2006): 49–75. Print.

Milstein, Tema. "Human Communication's Effects on Relationships with Animals." *Encyclopedia of Human-Animal Relationships: A Global Exploration of Our Connections with Animals.* Ed. Marc Bekoff. Vol. 3. Westport, CT: Greenwood Publishing Group, 2007. 1044–1054. Print.

———. "Nature Identification: The Power of Pointing and Naming." *Environmental Communication: A Journal of Culture and Nature* 5.1 (2011): 3–24. Print.

———. "'Somethin' Tells Me It's All Happening at the Zoo': Discourse, Power, and Conservationism." *Environmental Communication: A Journal of Nature and Culture* 3.1 (2009): 24–48. Print.

Morbello, M. "Zoo Veneers: Animals and Ethnic Crafts at the San Diego Zoo." *Communication Review* 1.4 (1996): 521–543. Print.

Plumwood, Val. "Androcentrism and Anthropocentricism: Parallels and Politics." *Ecofeminism: Women, Culture, Nature.* Ed. Karen J. Warren. Bloomington: Indiana UP, 1997. 327–355. Print.

———. *Feminism and the Mastery of Nature.* London: Routledge, 1993. Print.

Scott, James C. *Domination and the Arts of Resistance: Hidden Transcripts.* New Haven: Yale UP, 1990. Print.

Stibbe, Arran. "Language, Power, and the Social Construction of Animals." *Society and Animals* 9.2 (2001): 145–161. Print.

Wilson, Alexander. *The Culture of Nature: North American Landscape from Disney to the Exxon Valdez.* Cambridge, MA: Blackwell, 1992. Print.

Part III

Coherence

11 Listening with the Third Eye
A Phenomenological Ethnography of Animal Communicators

Susan Hafen

> I love dogs; it has always been clear to me that they lead extremely intense emotional lives. "No, Misha, no walk just now." What? *The ears would cock.* Can I have heard right? *"Sorry, Misha, but no." Unmistakable. The ears flop.... As he grew old, and could no longer walk as well, I could almost see him visit the scenes of his earlier life in his imagination.* (Masson and McCarthy xvi)

> *The white man has been only a short time in this country and knows very little about the animals; we have lived here thousands of years and were taught long ago by the animals themselves.* (a Carrier Indian qtd. in Abram 132)[1]

> *Who teacheth us more than the beasts of the earth, and maketh us wiser than the fowls of the heaven?* (Job 35:11)

INTRODUCTION: "THE LUNATIC FRINGE" OF ETHNOGRAPHY

This study began unofficially in 1994, when a graduate school friend told me about her sister Nancy who had attended workshops in California on human-animal telepathy. Not only could she communicate telepathically, but she could do so over the telephone, long distance, without any visual or auditory contact, and having never seen the animals. As a dog lover and inquiring skeptic, I was intrigued, especially since my friend prided herself on her scientific approach to knowledge. While my own experience, lived interpretively, persuaded me that animals and humans do indeed exchange intentional, nonverbal messages, Nancy's communication challenged my epistemological assumptions regarding perception, in particular, spatiality and corporeality. After several long telephone conversations, during which Nancy relayed information to me that my two Australian cattle dogs "told" or "showed" her—specific, detailed information she could not have known nor surmised—I decided that whatever communication phenomenon had occurred, I would someday study it.

Four years later, I attended a human-animal communication workshop and followed it up with phone interviews of the participants, resulting in a conference paper on an ethnography panel. Dwight Conquergood, one of our field's eminent ethnographers and the moderator of the panel, introduced my paper as "the lunatic fringe of ethnography." Response to the paper was mixed: Pet lovers were fascinated but skeptical about its "anthropomorphism"; hunters and meat eaters were suspicious that animal rights undergirded the topic; and two people contacted me to tell me animal communication stories of their own.

In 1998 few people had heard of animal communicators, or pet psychics as they are often called, although the terms are not interchangeable. Ethologists also write about their communication with animals, but typically do not refer to themselves as animal communicators. Although pet psychics call themselves animal communicators, some animal communicators dislike the term pet psychics. For Nancy, my friend's sister who appeared to communicate telepathically with my dogs, "psychic" is an umbrella term that also encompasses telling the future, reading tarot cards and/or receiving information via spirit guides. "Intuitive" is the term preferred by many animal communicators, writes Marta Williams in Learning Their Language: Intuitive Communication with Animals and Nature, because of the negative connotations associated with the term "psychic" (19–20). Others are comfortable with the term and call themselves pet psychics.[2] Penelope Smith, a pioneer in telepathic interspecies communication since 1971, calls herself an "animal mystic" (344). Today when "animal communicator" is Googled, the result is telepathic animal communication, and animal communicator/pet psychic services abound. Google displays over a million websites for "pet psychics" and "animal communicators." In the first ten pages of pet psychics on Google, fifty-eight people advertise their services (fifty-seven women, one man). At least three animal communicator certification programs are offered: the Assisi International Animal Institute, the Gurney Institute of Animal Communication and the Animal University. Besides advertisements, there are also testimonials and YouTube videos, as well as websites to discredit human-animal telepathy. In short, telepathic animal communication—which I will henceforth refer to simply as animal communication, the label used informally by its practitioners—has become a booming online business, as are seminars, workshops and certification programs.

The worldview espoused at these workshops decenters humans as the only sentient beings who communicate intentionally within and across species; thus, if humans can communicate telepathically, so might animals. This worldview is as legitimate as any other spiritual belief that can be neither proved nor disproved unqualifiedly through empirical replication (although both proponents and opponents might claim otherwise). The purpose of this study is to explore the phenomenon of human-animal communication through a detailed description of a workshop for its practitioners and, in so doing, to advocate for a more inclusive understanding of our deep, communicative bonds with other animals with whom we share the earth and our homes.

To understand human-animal communication from the standpoint of its practitioners (a loosely connected community), I have drawn upon Dell Hymes's ethnography of communication, which he extended beyond speech to "the full spectrum of modalities, or channels, employed in human communication" (16–17). Laminated onto this methodology is Merleau-Ponty's phenomenology, a "hermeneutics of ambiguity"[3] especially appropriate for this topic (Gallagher 3). This chapter is organized into three sections. The first is a rationale and explanation of my ethnographic methods, including a defense of "anthropomorphism." Second, I provide a description of the workshop activities and interactions. The third is an intertextual dialogue, drawn from follow-up interviews with the workshop participants interwoven with recent interviews of animal communicator educators and practitioners. This dialogue is then thematized with Merleau-Ponty's four categories of perception: relationality, corporeality, spatiality and temporality. The chapter ends with conclusions, caveats and a call to critical cultural communication scholars to affirm Bakhtin's statement that "the most intense and productive life of culture takes place on the boundaries" (2).

PHENOMENOLOGICAL ETHNOGRAPHIC METHODS: RATIONALE AND EXPLANATION

A study of telepathic human-animal communication may be the ultimate subaltern, outer edge of cultural boundaries in communication. It cannot be proven *true*, but, like religious and spiritual communication, it is *believed*; and that belief affects people's daily lives. A refusal to consider human communication with animals is tantamount to a refusal to take what Ruthellen Josselson calls the "empathic stance [which] orients us as researchers to other people's meaning-making, which is communicated to us through narrative" (32). Defining the empathic stance further, she explains that the point is to "interpret others who are themselves engaged in the process of interpreting themselves" (31). Said differently, my task in this study is to listen to animal communicators listening to themselves listen to animals. The listening practiced by human-animal communicators is a deeply intuitive, inward seeing—listening with the third eye, originally the All-Seeing Eye of the Goddess Maat of ancient Egypt.[4] Listening (and trying to see) as a workshop participant-observer is ethnography as embodied practice, which, James Clifford notes in *The Predicament of Culture*, "obliges its practitioners to experience, at a bodily as well as an intellectual level, the vicissitudes of translation" (24). Translation in this case meant understanding the beliefs, practices, intentions and struggles of people who wanted to become animal communicators.

Norman K. Denzin describes this kind of ethnography of communication as "interpretive interactionism," not only because it includes immersion in the phenomenon to be understood but also because the focus is on "those life experiences that radically alter and shape the meanings persons give to themselves and their life projects" (14). Max Van Manen differentiates hermeneutic phenomenology from ethnography in that it "does not aim to explicate meanings specific to particular cultures" (11) nor to offer "effective theory with which we can explain and/or control the world" (9) but instead aims to understand the essence and meaning of experiences as people live them. Perhaps the primary difference between ethnography and phenomenology is, in Van Manen's words, hermeneutic phenomenology's recognition that "life is fundamentally or ultimately mysterious" (16). However, both methodologies require interviewing, observing and using personal experience as a starting point in human communication research. Donal Carbough defines the ethnographer as someone who examines and re-creates moments of natural communication to reveal something general about the phenomena through the modes of "description, interpretation, and comparison"—and, in appropriate instances, criticism (262).

One communication mode for millions of people is telepathy, as reported in a 2005 Gallup poll in the United States and a 2006 Reader's Digest poll in Britain.[5] A few scientists have theorized the possibility of telepathy through other dimensions of reality, based on physicists' work on subatomic "nonlocality" (Schoen 165–174): Rupert Sheldrake proposes the existence of "morphic fields" (301–317); Michael Fox proposes an "empathosphere" (123). Given the credence of some scientists, it is not surprising that those who believe in telepathy and in animal communication might also extend telepathic communication to animals—creating a community of believers and practitioners. Indeed, there are likely many silent believers, fearful of being labeled one of the "lunatic fringe." An early 1978 book on animal telepathy was written by a research scientist under a pseudonym because he feared a prejudicial reaction to his anecdote-based investigations.[6] As Marc Bekoff says, "I'm a trained scientist. I'm not supposed to have these kind of universal connections—they put me in hot water with my associates. But I do" (qtd. in Schoen 172).

Bekoff, trained as an evolutionary biologist, now also calls himself an "anthrozoologist," an interdisciplinary addition to ethology, which is the study of animal behavior. Anthrozoology, explains Hal Herzog, cuts across the academic boundaries of psychology, anthropology, zoology, biology, sociology and ethology to "include the study of nearly all aspects of our interactions with other species" (16). Anthrozoologists such as Herzog now argue that anthropomorphism, attributing human characteristics to non-human beings, needs to be weighed against its opposite, anthropocentrism, the tendency to be human-centered, which is so natural that it is too often unquestioned. Leslie Irvine contrasts "critical/interpretive" with "sentimental" anthropormophism: The latter is a projection of human desires on animals without grounding interpretations of animal behavior on knowledge of the species and that particular individual in a specific context. In her book on dogs, Alexandra Horowitz says that anthropomorphism is necessary as the beginning of a hypothesis that still needs careful, operationalized observations. The trick is to "deconstruct" our anthropomorphisms to read canine behaviors from the dog's point of view (15). Animal communicators at this workshop, however, were reading dog's *minds* to understand their point of view.

This small, exploratory phenomenological ethnography focused on a community of (quasi) believers was initially based on attending a three-and-a-half-day workshop taught by Dr. Jeri Ryan[7] with thirteen participants (including me) and then conducting in-depth telephone interviews with ten of the participants. To those earlier observations of our experiences as participants and follow-up interviews, I have added more recent telephone interviews with four animal communicator teachers, two of whom have an international certification program, as well as three practicing animal communicators and one veterinarian who have attended multiple workshops for over a decade.[8] Interwoven into their accounts are quotes from writers on animal communication and animal behavior, thus creating an intertextual dialogue. Not represented here are the critics and skeptics of (telepathic) animal communication—not because their skepticism is unjustified, but because this is not an ethnography of their voices. All narrators gave permission to use their real first names and, in the case of the trainers, their full names.

HUMAN-ANIMAL COMMUNICATION WORKSHOP

The evening before the workshop, group members met Dr. Jeri Ryan at her introductory lecture, which included several activities to familiarize us with the practice of meditative listening. Individuals could pay a nominal fee to attend the lecture without paying the $350 to attend the three-day workshop, a cost that included continental breakfasts and vegetarian lunches. Jeri shared many of her own experiences communicating with animals—this was the last time she foregrounded her own experiences during the workshop because, as she would later explain, it was important that the participants concentrate on their experiences, rather than comparing themselves with the instructor. The first activity that evening consisted of selecting a button with a picture of an animal, gazing at the eyes of the animal and focusing on what that animal might have to say to us. Next, paired with someone we did not know, we gazed into his/her eyes and tried to sense something about that person. For some people, that exchange of perceptions was emotional and significant; for others, it was more mundane than revelatory. Jeri ended the evening by providing us with additional resources for study, books by animal communicators and "new age philosophers," whose views of reality include the interconnectedness of all life and make the phenomenon of human-animal telepathy seem plausible.[9]

On day one of the workshop, group members again practiced meditative breathing and did telepathic exercises wherein, eyes closed, they sent visual images to each other—for example, objects of a particular color. What we learned during these exercises is that some people are better senders of messages; others, better receivers. I was a better sender than receiver, unsurprising for an academic, but disappointing because I had hoped to hear my dogs' ideas, not merely share my own more clearly. At the end of the day, we each introduced our pets with whom we wished to communicate. Some people brought their cats in cages or had dogs on their leashes. Others, who had a longer distance to travel (including myself), brought pictures of our animal friends. Jeri made a promise to each of us that we would all, without exception, be able to communicate telepathically with an animal, and the way that we would know the information came from the animals and not our overactive imagination would be by verification with the animals' owners. She said that the largest barrier would be our own distrust in our abilities, and we were there to support one another. She was right; most of us did distrust the "messages" we received, regardless of the verification and support provided by others. We believed she could do it; we believed that others in the class might do it; we did not believe in ourselves. Our experiences were similar to those reported by journalist Arthur Meyers, who interviewed dozens of human-animal communication teachers and attended several workshops himself. He learned from his interviewees that telepathic communication is not a gift, but a skill that anyone can learn with sufficient motivation and practice and confidence (although some people seem to be more adept than others).

After each exercise, we went around the circle, talking individually about what we had sensed (or not). Several participants expressed their frustration that they did not seem to be "getting anything." Some had rich, detailed images but did not know whether the images were being "sent" by another person or were the product of vivid imaginations. Jeri urged "senders" to confirm even small aspects of images for the "receivers" to give us confidence that we were indeed communicating telepathically. For example, one participant sent another participant a red frisbee, the kind her dog played with. The receiver "saw" a red ball. When the sender disconfirmed her, saying "No, it was a Frisbee," Jeri intervened and said, "But what shape was it? What size?" Understanding, the sender then confirmed the receiver, telling her how close the image was—the shape and size were similar, and both objects were something the dog liked to play with.

We learned that part of seeing with our third eye was the expectation of seeing. As Fritjof Capra explains, the "observer" is in reality a "participator" whose expectations shape what is observed: "Natural science does not simply describe and explain nature; it is part of the interplay between nature and ourselves" (140). Merleau-Ponty says much the same thing: "The world is inseparable from the subject . . . and the subject is inseparable from the world, but from a world which it projects itself" (230).

In short, the difference between the more and less telepathically able participants was their beliefs in their abilities. Unlike classical science, where "seeing is believing," we needed to believe in order to see. The inseparability of the subject and the object—or of two subjects—is still science, but it is quantum science. The primary quantum uncertainty principle eliminates the "observer"; in physicist Heisenberg's words, "what we observe is not nature itself, but nature exposed to our questioning" (qtd. in Capra 140). In a participatory universe we are all intersubjective participators and unavoidably "entangled," as Marcus Chown explains in *The Quantum Zoo*: "What the observer [participator] knows is inseparable from what the observer is" (63).

The consequence for animal communicators is that belief in oneself—and the belief of others—is important in communicating with animals. In her *7 Steps to Communicating with Animals*, teacher Carol Gurney instructs beginners on how to deal with their doubts and to use supportive group members to validate the information they receive from animals (58–59). Another teacher, Marta Williams, explained in a recent telephone interview why an animal communicator might receive incorrect information when being challenged by skeptics: "Like test anxiety, with pressure, you might not do well, even though you have studied and at another time would have the answers." In other words, self-doubt and others' skepticism may (quantum) physically prevent the phenomenon.

On day two we began to work with the animals, both pictures and the live beings. We were given a list of "getting to know you" questions to ask the animals, if we so chose, such as:

> What do you like most/least about your life?
> How do you feel about your environment/humans/other members of your species?
> How is your health?
> What makes you proud?
> What would you like to ask/tell me?

Many group members soon discovered that it was easier to work with the pictures than with the animals themselves, sitting across the room next to their owners. Concentrating on communicating with a being who appeared to be oblivious to you weakened your belief in yourself. Wrinkling her forehead in concentration, one participant closed her eyes and mentally tried to reach the cat across the room. The rest of us watched. The cat rolled over—not a wink, a glance, a paw or a whisker pointed in the person's direction. If the cat could "hear," wouldn't he give her some sign of recognition? Not necessarily, said Jeri. Visually disconfirmed, the participant gave up on the cat. Other participants, however, did receive communication from animals present; and their owners confirmed the information, ignoring the animals' apparent, nonverbal disregard.

Information received varied from descriptions of their favorite toys and activities to images of their backyards or sleeping places. Emotions were almost always present: Animals expressed jealousy of other pets, loneliness for their owners, love and gratitude for their homes, complaints about inattention, identifiable aches and pains or appreciation at being contacted. One of the group members, an acupuncturist, was able to pinpoint pain in a horse's shoulder, of which her owner was unaware but, upon reflection, remembered an incident that might have caused an injury. The acupuncturist was one of the group's best "receivers" of information; however, she was dissatisfied with the kinds of information other participants supplied about her own dogs and cats at home—none of it seemed to fit. One of the group members was a scientist—a biologist known for her no-nonsense, objective and detached approach. Another participant was a university administrator, who disdained anything "new age." Although both expressed doubts about their abilities, other participants continually confirmed their reports. Indeed, I was envious of the administrator's "tingling" sensations when she received distinct communication (sensations most of us did not experience) and the biologist's olfactory perceptions (she smelled aspects of the animal's environment during communication).

Initially skeptical, I was surprised at the detailed, accurate descriptions several group members gave me about conversations they had with my dogs. In one activity we were asked to contact the animal by seeing it in our mind's eye, saying hello, touching it in some way and then waiting for a response. The person who selected my Australian cattle dog Bronte to contact reported that at first nothing had happened, then Bronte's head appeared; she stretched her neck and opened her mouth wide, in a long, exaggerated yawn, looked rather critically at the participant, but remained silent. I was stunned. Bronte's propensity for exaggerated yawns at all times of the day was a distinguishing trait, and her lack of desire to communicate unless she wanted something was also typical. Such a description for my other dog would have been inconceivable. Later, another participant selected my older cattle dog, Toklas (mother to Bronte), and reported Toklas's irritation with Bronte, "who thinks she's so tough" (true!), but also reported that she was "beautiful" (my oft-used word for Bronte). I was convinced that both group members, looking at photographs and then "bringing [their] energy from [their] heart to [their] third eye, connecting [their] hearts, [their] intention, and [their] minds"— according to Jeri's instructional handout—had somehow contacted my dogs. By the end of the second day, we had all witnessed what appeared to be human-animal telepathy, even if we weren't confident of our own abilities as senders or receivers.

Day three was the most demanding, requiring the greatest suspension of disbelief. Despite some reservations, we all participated thoughtfully with day three's shamanistic journey with our totem, power animals. This dyadic exercise was designed to help us learn from an animal teacher. One partner of the dyad acted as "shaman" to the other partner, who silently formulated a question about his or her life. The shaman, eyes closed, listening to a drumming beat (Jeri's audiotape), visualized a journey accompanied by one or more animals, as if in a dream state. At the end of the drumming, the shaman, now "awake," described the journey to his or her partner. The partner then tried to interpret the journey as a metaphorical answer to his or her question.

The secret to metaphorical thinking is that we find in the metaphors the answers we already know, albeit answers we may not know that we know. For some partners, the metaphorical journeys represented profound answers to their questions. Others needed more time to ponder the metaphors. The animals who led the shamans on the journey were symbols in our lives, their meanings explicated in various textbooks that Jeri shared with us.[10] For example, one participant's power animal was a squirrel, which can mean "you are being told to honor your future by readying yourself for change" (Andrews 141). My shaman was escorted on a journey by frogs— apparently my "totem/power" animal for that time in my life. My reaction was disappointment—not a wolf or an eagle, but a frog? Reading from one textbook, however, I learned that frogs are associated with heightened sensitivity to the environment and emotions, something I recognized about myself. Much later, upon returning home, I noticed for the first time just how many frogs are in my house—gifts from friends, who know I like unique, idiosyncratic images of animals. Frogs sat on my stereo, window sill and plant holders. Because I had not bought them myself and have no particular attachment to frogs, I had never noticed their accumulation—a fact that startled me. Prior to the workshop I did not know the woman who acted as my shaman, so how had she known? More importantly, how had I *not* known?

If some group members' initial reactions to the first shamanistic journey were mixed, their reactions to the second journey were more incredulous: With different dyadic partners, we took shamanistic journeys to a past life of an animal companion of our partner. Discussing past lives in general prior to the activity, not all participants believed they had lived past lives, although this is a common premise of animal communicator books.[11] Even fewer had contemplated the past lives of their pets. Marta Williams acknowledges in her chapter "When They Return" that "concepts of spirituality and reincarnation in animals are difficult for many people to accept," but, like religious beliefs, they can give comfort without proof, especially in coping with the unexpected death of an animal (*Beyond Words* 127–148). Still, trying to suspend disbelief, we embarked on journeys that provided some surprising insights into our animals. Bronte, for example, had been a moose in a past life—an image very congruent with her predisposition to contemplate a visitor silently, then charge without warning. This excursion into the outer boundaries of possibilities became another way to connect with the animals we loved, whether metaphorically or spiritually.

At the end of the workshop, despite being emotionally drained from the concentrated mental focus required, most group members seemed committed to practice what they had learned. Many expressed the desire to stay in contact with the other participants with whom they had shared emotional experiences, so one participant volunteered to set up an Internet group mailing list. Participants within driving distance of one another agreed to try to meet monthly to practice their human-animal telepathy. Not living in the area, I was unable to meet with them, but for over a year I received email announcements of their meetings and messages from people who did not attend the workshop but who joined the group online. For example, four months after the workshop, one group member emailed us with a request: Was there a way to communicate telepathically with wild animals to tell them to avoid highways? This person had seen too many dead deer, raccoons and other woods creatures on the roads and wondered if we could help. Several members responded that they would attempt to send warning messages. Whether or not they made the attempt, and whether their messages were received, the point—which Jeri Ryan reiterated in a recent telephone interview—is the community the workshop created.

FOLLOW-UP INTERVIEWS WITH WORKSHOP PARTICIPANTS

Participants in the workshop were not pet "owners"; rather, we were the friends, companions, caretakers and guardians of certain animals in our lives. Indeed, "pet owner" is viewed as so demeaning to animals who are, in essence, family members that nearly twenty cities (primarily in California) and the state of Rhode Island use "guardian" rather than "owner" in their animal control ordinances.[12] Penelope Smith, a pioneer in telepathic interspecies communication since 1971, states that she "doesn't believe we ever own animals" because ownership does not "honor the sense that they have their own lives and their own paths" (126). Another view comes from animal behaviorist Patricia McConnell, who writes that although "owner" is awkward—we don't own our children—so is "guardian" (xiv). Further, "my pet" can be a term of endearment, as journalist P. Elizabeth Anderson notes, but Anderson goes on to say that she does not use the phrase publicly because words have power and "we need to be nudged toward a greater appreciation for the lives of animals" (xix). Whatever we called our dogs, cats, horses or birds, we hoped to increase our communication with them and so were (somewhat) open to more radical ideas about personal transformation—such as developing telepathic abilities. I have used both "owner" and "pet" in this chapter to describe groups of people who do legally own their animals, which—by virtue of providing emotional rather than merely functional connection—can be termed "pets." However, the point that Penelope Smith and animal communicators make about language is an important one that signifies their relationship to beings who are, as animal communication teacher Joyce Leake said in a telephone interview, "not *just* animals." Jeri Ryan clarified this: "Animals have their own purposes and goals, and a culture and DNA that guide their functions, just like we do." No one in this workshop thought otherwise, regardless of the words used.

What I hoped to learn from interviewing workshop participants was (1) how group members' relationships with their animals related to human-animal telepathy, and (2) what impact the workshop itself had on their life and their interactions with animals. Approximately one month after the workshop, I conducted one-to-two-hour telephone interviews with ten of the group members and asked the following:

(1) Describe the role or relationship your pets have had in your life.
(2) Describe your feelings toward your pets and how you communicate those feelings.
(3) How do you know that your pets understand your feelings—that is, receive your messages? How do they reciprocate?
(4) What feelings or messages do your pets communicate to you? How do you know that they are communicating those feelings or messages?
(5) How do your pets communicate with the environment by their interactions?
(6) What would you like to learn from your pet(s)? How could it teach you? Or how could you learn?
(7) Has the workshop changed any part of your life? If so, how?

These seven questions were designed to elicit responses that might create an embodied, relational-centric understanding of the phenomenon of human-animal communication, based on Merleau-Ponty's view of perception:

> In the perception of the other, this happens when the other organism, instead of behaving like me, engages with the things in my world in a style that is at first mysterious to me but which at least seems to me a coherent style because it responds to certain possibilities which fringed the things in my world. (qtd. in Yeo 48)

Although Merleau-Ponty is not specifically describing animals as "the other organism," he might well be, given his attunement to nature and his belief in Jakob von Uexküll's *Umwelt*, which "provides a convenient starting point for connecting Merleau-Ponty's ecological insights with bio-semiotics" (Harney 135). Coined in 1926, the term *Umwelt* refers to the perceptual world of one's physical environment, which is species specific, framed as the "web of semiotic relations centered on the cognitive organism" (Deely 98). Merleau-Ponty is referring to the shared *Umwelt* of all animals (including humans) when he writes,

Why would not the synergy exist among different organisms, if it is not possible within each? Their landscapes interweave, their actions and their passions fit together exactly: This is possible as soon as we no longer make belongingness to one same "consciousness" the primordial definition of sensibility . . . (*Visible* 142)

In *Merleau-Ponty and the Generation of Animals,* Bryan Smyth admits that although Merleau-Ponty's phenomenology was humanist in nature, it was decidedly a nonanthropocentric humanism with biocentric sensibilities grounded in what Merleau-Ponty called "interanimality." The authors of the anthology *Merleau-Ponty and Environmental Philosophy* explore his philosophy for themes concerning "the intertwining and mutual well-being of all life forms, questions about the meaning of being human, and the refusal to value Nature solely in terms of its potential for human use" (Cataldi and Hamrick 5). One theme is Merleau-Ponty's four categories of perception—relationality, corporeality, spatiality and temporality—which can be used to situate the first six questions asked of workshop participants.

Relationality is "the lived relation we maintain with others in the interpersonal space that we share with them," permitting "a conversational relation which allows us to transcend our *selves*," which ultimately leads to "the communal, the social for a sense of purpose in life . . . (Van Manen 105). Questions one and two address the lived relations people had with their animals. Roles participants named included pets perceived as children (or grandchildren, in the case of one participant's mother), outdoor companions, protectors, comforters, friends and an avenue to "escapes." For example, Donna said that, when feeding her horses, she feels "like I'm on vacation, relaxed." The variety of these roles is similar to those that human companions play in our lives. In my case, one of my dogs relates to me as a wise friend; the other, as a demanding child. When talking about their feelings toward their pets, Jamie and Curtis, a married couple who interviewed in tandem on the telephone, tried to communicate that they consider their cats equals—"for example, we tell them we're going to work and what time we'll be back." Almost all (nine) participants named love as one of the feelings they communicated, both verbally and/or with physical touch. In *Dog Love*, Marjorie Garber recounts how expressing love for animals (petophilia) has been viewed as a "moral danger," "compensatory substitute" or even "erotics of dominance." She retorts to critics of animal relationality:

The point is perhaps not to argue about whether dog [pet] love is a substitute for human love, but rather to detach the notion of "substitute" from its presumed inferiority to a "real thing." Don't all loves function, in a sense, within a chain of substitutions? For Freud, the original object is the parent, and all nonincestuous loves are then substitutes for the love that has become taboo. . . . We learn to love by loving, and it is a long, indeed, an interminable process. To distinguish between primary and substitute loves is to understand little about the complexity of human emotions. (Garber 135)

Participants also talked aloud to their pets, touched them, massaged them, slept with them, gave them special treats, celebrated their birthdays and hung their Christmas stockings. These intimate relationships are economically important, as documented in Michael Schaffer's *One Nation under Dog*: $41 billion a year in 2007 with a projected increase of seven percent growth each year, proving the pet market to be recession proof (15–16). Whether or not all of the extravagances bought for what Schaffer calls "fur babies" are actually items the animals want versus what the people want (hamster mineral water?), the point is the value of these relationships in a capitalist society. Marc Bekoff, in *The Emotional Lives of Animals*, explains the billion-dollar pet industry as the result of "shared emotions and their gluelike power to attract and bind" (19).

Corporeality is the way in which we experience others as utterly separate yet physically close because "we are always bodily in the world" (Van Manen 103). Question three asked participants to describe how they know their pets understand their feelings/messages and how their pets reciprocate in communicating feelings/messages. This is received knowledge experienced bodily. Jennifer G. said that her cats could read her facial expression, and Donna said that her tone of voice was important, especially a happy voice, in getting her cats' attention. Most based their knowledge that their pets understood them on the experience of looking into their pets' eyes and seeing what they described as the eyes "lighting up." Animal behaviorist Patricia McConnell says that when reading emotions in both dogs and humans, the "hard eye" signals danger and the "warm eye" signals a warm heart (58–60). In a telephone interview, veterinarian Nan Larsen contrasted dogs with cats, who don't use their mouths as much as dogs. When working with fearful or mean cats, she will blink her eyes slowly and gently, which tells the cat that she is not threatening.

There were many more examples of animals responding to their owners' verbal communication: Gloria's Schnauzer running to the kitchen when asked if she wants a treat, Barb's Dalmatian halting reluctantly when commanded not to chase the cat and Vicki's Australian Shepherd's immediate submission when chastised. Beyond verbal communication, Pam and Jennifer G. told how their pets read their minds. Describing her cats, Jennifer G. explained, "I'll be thinking of doing something for them, giving them something to eat, and they'll start to respond by moving or going towards that thing—they'll jump off my legs and go in that direction." Pam was emphatic about her horses: "They read my every thought." Workshop participants named bodily movements such as wagging tails, licking and paws placed on the owners; these behaviors seemed to indicate that their pets had received telepathic communication. Still, for most participants, unequivocal telepathic communication with their animals was more a goal than a reality. Animal communicator Marta Williams suggested that the best way to have your animals respond to you is to talk to them aloud, often, explaining things to them; do this for two weeks and see if they don't change their behavior.

Question four asked about feelings and messages *from* their pets and how participants knew that the animals were actually communicating those messages. The most frequent messages were "feed me" and "let me out," shown by the animals leading them to the door or to their food bowls. Other messages were the communication of feelings, which included disappointment, irritation, worry, pain and happiness, which participants read into their faces and "just knew." However, most participants, like me, seemed to be better senders than receivers, particularly with their own animals. Books by animal communicators all give the same advice: Trust yourself, pay attention to the information you receive and believe that it comes from the animals and not your own imagination.[13] That does not mean, however, that you are always correct.

When I recently asked seven animal communicators, as practicing professionals, if they were sometimes incorrect about what the animals had apparently told them, they all said yes.[14] What went wrong? Julie Morgan said that she doesn't necessarily get the wrong information, but her interpretation might be off: "Like yesterday when a cat wanted to have the litter box to herself, I thought that there were other cats using it, but it turned out that the dog was using it. It can be like translating French into English." Paula Davis said, "It's not an exact science; it's easy to pick up on someone else. So if you are in a home with two dogs, the other one can butt in and can give you images." Joyce Leake said that the communicators themselves might not be in tune; they may become distracted and need to refocus or, Carol Gurney added, "you just may not be batting 100% that day, everyone has off days" (including psychotherapists, I might add). Jeri Ryan noted that animals, like humans, also change their mind: "Just because a dog says he won't bite, I would never try to predict that he won't bite." The best way to know that it is the animal and not your imagination is to verify the information; however, "sometimes you're wrong because the person doesn't know the right answer," stated Marta Williams. An example might be when a dog shows you how she happily chases quail, but the person says, no, there aren't any quail around here; later, the person might call you and say, guess what, there are quail, and my dog chased after them. Nancy Gardner said that it takes time to distinguish between your mind and the animal's mind. Two clues to watch for, she said, are when something—seen through the animal's eyes—flashes in your mind very quickly or when the information is something you would never have thought of; when these things occur, the information is usually coming from the animal's mind.

Spatiality refers to "the world or landscape in which [we] move and find [ourselves] at home. . . . a category for inquiring into the ways we experience the affairs of our day to day existence" (Van Manen 102–103). Question five asked how their pets communicated with the environment (outside the house) through their interactions; all ten participants mentioned ways that their animals rejoiced in the natural world, with some differences and qualifications. All the dogs liked to go on walks: Sharon's Shitzu and Great Pyrenees assume "this is an adventure and everyone will be friendly" and are never aggressive with people or animals. On the other hand, Gloria's Schnauzer has to be restrained from people in uniforms. Many cats and dogs were protective of their space, guarding their yards, but like people, showed individual differences. Jamie and Curtis have one cat who likes to go out, especially to smell flowers, and another cat who is too afraid of being abandoned and sudden noises (e.g., lawn mowers, sprinklers) to stay in the yard alone. Barb has one cat who hunts fearlessly (mice, birds) and another who is too afraid of being hunted herself by Barb's dogs. Jamie's Australian Shepherd, trained for search and rescue, has learned to ignore any part of the environment that does not relate to the victims; her Great Pyrenees, on the other hand, will get off track to follow a deer trail or watch a bird. As participants described how their animals related to the external environment, many described ways in which they shared it together, shaping the way they experienced the world. This shared love of the natural world between human and their animal friends is what E. O. Wilson calls "biophilia," or the "[innate] urge to affiliate with other forms of life" (85). Wilson believes this affiliation is based on the "phylogenetic continuity of life" (130) between animals and humans. He adds that "within the matrix of humans and other organisms" (139) resides our love of nature—the naturalist is within us all (although more deeply rooted in some than others).

Temporality is lived time, "subjective opposed to clock time or objective time . . . dimensions of past, present and future . . . the way I carry myself (hopeful or confident, defeated or worn-out) . . . a perspective on life to come" (Van Manen 104). Question six addressed what pets could teach the participants about living in the world. Implicit in this question is a sharing of each species' biophilia. Allen Schoen says that biophilia has "served us well" by allowing us to form bonds with animals who have helped us to be more successful hunters, gatherers and foragers (164–165). Abram adds another advantage of our connections with other species—if we pay attention to them, we gain the knowledge of "these Others," who are the "carriers of intelligence we often need" (14). He continues with examples: Animals inform us of unseasonable changes in the weather, imminent eruptions, where the ripest berries are found and sometimes how to find our way back home.

Workshop participants learned from their animals how to pay more attention to the natural world and be in the present. "Living in the present" means living in harmony with nature for Jennifer B., Donna and Gloria; it means "not fearing death" for Jamie and Jennifer B. Pam and Gloria both noticed their increased awareness of smells. Jennifer G. learned patience while watching her cats watching robins pluck worms out of the ground. Riding horses, Donna has become more aware of the ground—its wetness, rockiness or slipperiness—whether she is on or off her horse; she "just feels more alive—*breathes*." Jennifer B. sees birds more because her cats are constantly scanning the skies. With her dogs, Jamie feels more like a kid. Watching his cats explore their small yard, Curtis dreams of a future when he can "have more land, more space." Sharon has found that whenever she meets any animals, "there's a stronger and quicker reaction [from them]. I love that." She continued, "I am drawn to notice what they notice . . . makes me slow down and take time—it's very peaceful."

When asked what they hoped to learn from their pets, both Donna and Pam described their desire to be in harmony with nature, to accept it (Sharon added that this can mean "not pruning"). Gloria wants to "be more aware of the immediate environment by paying attention to [her Schnauzer]; likewise, Sharon wants to see "through their eyes," and Barb wants to "know what they smell." Jennifer G. wants not only to see nature as it is, but to see the animals for "*who* they are." Gloria believes that animals can teach her how to communicate better emotionally; Barb and Jennifer G., to give unconditional love. Pam and Curtis hope that more two-way communication with animals might help humans get along better not only with different species, but also with different races of their own species.

Professional animal communicators hope for even greater lessons from communicating with our animals—extending temporality from this life to the next. Books by Marta Williams, Penelope Smith and Dawn Baumann Brunke all provide examples of their connections with deceased or reincarnated animals, as well as advice for how to make these connections. Although no workshop activities involved contacting animals who had died, Jeri Ryan told us that communication was indeed possible. Listening with one's third eye to animals could shape our experience of relationality, corporeality, spatiality and temporality in ways we had not imagined. What we could imagine were ways that this workshop might change our lives, and that was the subject of the last question for the participants.

Many participants talked about their renewed goals to strengthen their communication skills with animals. Jamie had read more books about animal-human communication to help her. Pam, who described herself as "intuitive" since childhood, said that communication for her goes beyond animals to "wind, air, earth, plant life"; she believed her ability to communicate with the environment continued to grow. Donna had an increased sense of the equality of animals with humans, based on the respect she saw in other participants. Vicki too had developed a greater sense of equality. She had always believed animals had spirits, but she had thought of them as a lower form: "Now I see it as the form they took in this life." Curtis, too, found that the workshop "elevated animals to more than animals. It let me into *their* world, rather than just bringing them into our world."

Some participants pondered how respecting animals would affect their practices. Gloria, the biologist who had to kill salamanders for research, said that she would now "stop to think about how the salamander would respond to a situation." Barb, the acupuncturist, wanted to use more conscious communication when using acupuncture with animals so she "can decipher problems and work with them more effectively." Jennifer B. believed that, if she practices, "I'll get better at it . . . I could do it professionally." Jennifer G., the only participant who had previously attended human-animal communication workshops, was committed to continuing her skill in animal-human communication, and hoped to work in a shelter or with therapy animals someday. Her goal is greater spiritual awareness, a goal she wishes everyone will achieve: "If people experienced this kind of relationship with animals, they would have to adjust their view of the environment—seeing more life and spirit in all of plants and all of nature." Merleau-Ponty would agree: "the becoming-nature of man [and woman] . . . is the becoming-man of nature" (*Visible* 285).

(NEVER FINAL) CONCLUSIONS, A CALL FOR RESEARCH AND CAVEATS

"Becoming" requires listening, not a Western cultural strength. Deep listening is associated with highly contextualized Eastern cultures, whose religions—such as Hinduism and its third eye—retained what scientist Robert Boyle called in the late seventeenth century "a discouraging impediment to the empire of man over the inferior creatures" (qtd. in Thomas 22). Listening is also more commonly associated with native peoples in oral cultures; the Greek alphabet, according to Abram, "effectively severed all ties between the written letters and the sensible world from which they were derived" (111). Abram further suggests that what we have lost is the sensual psyche via "a language that has forgotten its expressive depths" (85). Animals, whose entire bodies quiver with expression, can help us to "return to things themselves . . ." or, in Merleau-Ponty's words, "that world which precedes knowledge, of which knowledge always *speaks* . . ." (*Phenomenology* ix). That world is Nature. Instead of wresting it from the animals, we need their help in valuing nature as they do.

This phenomenological ethnography of communication, focused on one group of people interacting at a three-and-one-half-day workshop, is just one example of the contribution that communication scholars could make to the larger endeavor of changing our relationship with nature from one of control and dominance to one of cooperation and respect. Ethnographic and phenomenological projects can have transformative effects on both interviewer and participants, the result of "deep learning, leading to a transformation of consciousness, heightened perceptiveness, increased thoughtfulness and tact, and so on" (Van Manen

163). Research that not only contributes knowledge about lived experience, but also transforms us in the process of gaining that knowledge is research using the third eye. Transformation based on a rearticulation of power is the goal for ethnography situated as "critical cultural politics," according to Dwight Conquergood. Despite his characterization of human-animal communication as the "lunatic fringe" in ethnography, this study addresses his own questions:

> What kinds of knowledge, and their attendant discursive styles, get privileged, legitimated, or displaced? How does knowledge about communication get constructed? What counts as an interesting question about human communication? . . . And, most importantly for critical theorists, what configuration of socio-political interests does communication scholarship serve? How does professionally authorized knowledge about communication articulate with relations of power? (Conquergood 193)

What has served as "knowledge" has been vested in the power relations of human domination of animals—a relationship founded upon the need to objectify beings we use as labor or harvest for food or material goods. That objectification has continued despite centuries of stories, novels and (auto) biographies testifying to the importance of animal relationships in human lives. Communication scholars might not be qualified to testify on the question of animals' mental and emotional abilities, but we can represent the stories humans tell themselves and others about their communicative relationships with their pets, their animal companions. The legitimacy of spiritual and religious communication is not based on first proving God exists, nor should the legitimacy of human-animal communication be based on the extent of its mutuality. Animal communicators have pushed the envelope further by asking why, if humans can communicate emotions and meanings telepathically (a widespread belief), animals should not be able to do the same. Perhaps animals' limitations in using symbolic language might make them more adept at sending mental pictures, as they are more adept than humans in other capacities.

Animal research in the past has been characterized by what Jeffrey Masson calls a "religious" rather than "logical" prejudice toward experimental versus anecdotal evidence, despite the myriad of published anecdotes that have served as evidence in the social sciences (16–19). Today, as scientists begin to recognize and respond to animals as thinking, feeling subjects, rather than instinct-bound objects, increasingly shrugging off accusations of anthropomorphism, moral and ethical issues will increase.[15] The worldview of animal communicators, by embracing alternative communication modes such as telepathy and decentering humans as communicators, results in a broader and more inclusive model of communication. It also begins to shift our relationship to the earth itself as well as our relationships with all sentient beings and, ultimately, with ourselves as sensing, intuiting, human animals.

NOTES

1. In The Spell of the Sensuous: Perception and Language in a More-Than-Human World (New York: Pantheon Books, 1996), David Abram describes the many things we have learned from animals. Indigenous peoples have traditionally shown more respect for animals as teachers who could help them "read the earth," learning which plants are edible or useful as medicine, where to find water and when storms are coming. Indeed, for the Inuit (and numerous other peoples), humans and animals originally spoke a common language.

2. In phone interviews, animal communicators Paula Davis (December 3, 2011) and Joyce Leake (December 5, 2011) both said that they accepted either "pet psychic" or "animal communicator"—whichever one people felt comfortable using.

3. This term is used by Shaun Gallagher, "Introduction: The Hermeneutics of Ambiguity," Merleau-Ponty, Hermeneutics, and Postmodernism, ed. Thomas W. Busch and Shaun Gallagher (New York: State U of New York P, 1992), 3.

4. A brief explanation of the "All-Seeing Eye" is provided by Barbara G. Walker, The Woman's Encyclopedia of Myths and Secrets (San Francisco: Harper & Row, 1983), 294–295. The third eye chakra of Hinduism is described by Diane Stein, Stroking the Python: Women's Psychic Lives (St. Paul, MN: Llewellyn, 1989), 203–230.

5. For information on the American Gallup poll, see David W. Moore, "Three in Four Americans Believe in Paranormal," Gallup News Service 16 Jun. 2005 <http://home.sandiego.edu/~baber/logic/gallup.html>. For information on the British Reader's Digest poll, see "Most Britons Accept ESP, Survey Reveals," UK Psychics Report 6 Jun. 2006 <http://www.ukpsychics.com/paranormal_survey.html>.

6. Joseph Wylder (pseudonym), Psychic Pets: The Secret Life of Animals (New York: Gramercy Books, 1978). He refers frequently to an earlier book on human-animal telepathy by J. Allen Boone, Kinship with All Life (San Francisco: Harper & Row, 1976).

7. Dr. Jeri Ryan is a licensed clinical psychologist and one of the founders of the Assisi International Animal Institute (AIAI) in Oakland, California, a nonprofit educational and rehabilitation center. Organized in 1995, it is dedicated "to promoting the increased well being of all non-human animals and the enhancement of all relationships of humans with non-human animals" (http://www.assisianimals.org/mission.html).

8. All telephone interviews were conducted between November 29, 2011, and December 9, 2011, with the following people, who have given permission for me to use their first and last names: Carol Gurney, animal communicator teacher, author and founder of The Gurney Institute for Animal Communication; Marta Williams, animal communicator teacher and author; Jeri Ryan, animal communicator teacher and founder of Assisi International Animal Institute; Joyce Leake, animal communicator teacher, author and founder of Animal University; Paula Davis, animal communicator; Nancy Gardner, animal communicator; Nan Larsen, veterinarian and owner of Dancing Cats Feline Health Center; and Julie Morgan, animal communicator.

9. In particular, Ryan recommended books by animal communicator Penelope Smith, Animal Talk: Interspecies Telepathic Communication (Point Reyes Station, CA: Pegasus, 1989) and Animals: Our Return to Wholeness (Point Reyes Station, CA: Pegasus, 1993). She also recommended Fritjof Capra's The Tao of Physics (see Works Cited).

10. Two books we used: one by Ted Andrews (see Works Cited) and one by Jamie Sons and David Carson, *Medicine Cards: The Discovery of Power Through the Ways of Animals* (Santa Fe, NM: Bear & Company, 1988). Another recommended book is by Jessica Dawn Palmer, *Animal Wisdom: The Definitive Guide to the Myth, Folklore, and Medicine Power of Animals* (London: Thorsons, 2001).

11. See Penelope Smith (221–264) and Dawn Baumann Brunke (see Works Cited); see also Jean Houston, *Mystical Dogs: Animals as Guides to Our Inner Life* (Makawao, HI: Inner Ocean Publishing, 2002).

12. Herzog (74). Also, dvm newsmagazine online (January 1, 2011) lists the specific municipalities that have changed "pet owner" to "guardian": the California cities of Beverly Hills, San Jose, Imperial Beach, Santa Clara County, Albany, Sebastopol, Marin County, San Francisco, West Hollywood and Berkeley; Bloomington, Indiana; St. Louis, Missouri; Wanaque, New Jersey.; Woodstock, New York; Amherst, Massachusetts.; Menomonee Falls, Wisconsin; Sherwood, Arkansas; Boulder, Colorado; and Windsor, Ontario, Canada.

13. Read Gurney's seven-step Heartfelt program (9–34). See the exercises for "better reception" in Williams, *Learning* (117–135).

14. See note 8 for information on animal communicators.

15. For a thorough discussion of the recuperation of animals as subjects, see Jonathon Balcombe, *Second Nature: The Inner Lives of Animals* (New York: Palgrave MacMillan, 2010); and Marc Bekoff and Jessica Pierce, *Wild Justice: The Moral Lives of Animals* (Chicago: U of Chicago P, 2009). See also Aaron Katcher and Alan M. Beck, eds., *New Perspectives on Our Lives with Companion Animals* (Philadelphia: U of Pennsylvania P, 1983); and James Serpell, *In the Company of Animals: A Study of Human-Animal Relationships* (New York: Basil Blackwell, 1986).

REFERENCES

Abram, David. *The Spell of the Sensuous: Perception and Language in a More-Than-Human World.* New York: Pantheon Books, 1996. Print.

Anderson, Elizabeth P. *The Powerful Bond between People and Pets.* Westport, CT: Praeger Publishers, 2008. Print.

Andrews, Ted. *Animal-Speak: The Spiritual and Magical Powers of Creatures Great and Small.* St. Paul, MN: Llewelyn, 1996. Print.

Bakhtin, Mikhail. *Speech Genres and Other Late Essays.* Trans. Vern McGee. Ed. Caryl Emerson and Michael Holquist. Austin: University of Texas Press, 1986. Print.

Bekoff, Marc. *The Emotional Lives of Animals.* Novato, CA: New World Library, 2007. Print.

Brumke, Dawn Baumann. *Animal Voices, Animal Guides: Discover Your Deeper Self through Communication with Animals.* Rochester, VT: Bear & Company, 2009. Print.

Chown, Marcus. *The Quantum Zoo: A Tourist's Guide to the Neverending Universe.* Washington, D.C.: Joseph Henry Press, 2006. Print.

Capra, Fritjof. *The Tao of Physics: An Exploration of the Parallels between Modern Physics and Eastern Mysticism.* 3rd ed. Boston: Shambhala, 1991. Print.

Carbaugh, Donal. "The Critical Voice of Ethnography in Communication Research." *Research on Language and Social Interaction* 23 (1989/1990): 261–282. Print.

Cataldi, Suzanne L., and William S. Hamrick. Introduction. *Merleau-Ponty and Environmental Philosophy: Dwelling on Landscapes of Thought*. Ed. Suzanne L. Cataldi and William S. Hamrick. Albany: State U of New York P. 1–15. Print.

Conquergood, Dwight. "Rethinking Critical Ethnography: Towards a Critical Cultural Politics." *Communication Monographs* 58 (1991): 180–196. Print.

Clifford, James. *The Predicament of Culture*. Cambridge: Harvard UP, 1988. Print.

Davis, Paula. Telephone interview. 3 Dec. 2011.

Deely, John. *Introducing Semiotic: Its History and Doctrine*. Bloomington: Indiana UP, 1982. Print.

Denzin, Norman K. *Interpretive Interactionism*. Applied Social Research Methods Series, 16. Newbury Park, CA: Sage, 1989. Print.

Fox, Michael. "Interrelationships between Mental and Physical Health: The Mind-Body Connection." *Mental Health and Well-Being in Animals*. Ed. Franklin D. McMillan. Ames, IA: Blackwell, 2005. 113–126. Print.

Garber, Marjorie. *Dog Love*. New York: Simon & Schuster, 1996. Print.

Gardner, Nancy. Telephone interview. 29 Nov. 2011.

Gurney, Carol. Telephone interview. 30 Nov. 2011.

———. *The Language of Animals: 7 Steps to Communicating with Animals*. New York: Bantam Dell Publishing, 2001. Print.

Harney, Maurita. "Merleau-Ponty, Ecology, and Biosemiotics." *Merleau-Ponty and Environmental Philosophy: Dwelling on Landscapes of Thought*. Ed. Suzanne L. Cataldi and William S. Hamrick. Albany: State U of New York P. 133–148. Print.

Herzog, Hal. *Some We Love, Some We Hate, Some We Eat*. New York: Harper Perennial, 2010. Print.

Horowitz, Alexandra. *Inside of a Dog: What Dogs See, Smell, and Know*. New York: Scribner, 2009. Print.

Hymes, Dell. "The Anthropology of Communication." *Human Communication Theory: Original Essays*. Ed. Frank E. S. Dance. New York: Holt, Rinehart, and Winston, 1967. Print.

Irvine, Leslie. *If You Tame Me: Understanding Our Connection with Animals*. Philadelphia: Temple UP, 2004. Print.

Josselson, Ruthellen. "Imaging the Real: Empathy, Narrative, and the Dialogic Self." *Interpreting Experience: The Narrative Study of Lives*. Ed. Ruthellen Josselson and Amia Lieblich. Thousand Oaks, CA: Sage, 1995. Print.

Larsen, Nan. Telephone interview. 29 Nov. 2011.

Leake, Joyce. Telephone interview. 5 Dec. 2011.

Masson, Jeffrey Moussaieff. *Dogs Never Lie about Love: Reflections on the Emotional World of Dogs*. New York: Delta, 1995. Print.

Masson, Jeffrey Moussaieff, and Susan McCarthy. *When Elephants Weep: The Emotional Lives of Animals*. New York: Delta, 1995. Print.

McConnell, Patricia. *For the Love of a Dog: Understanding Emotion in You and Your Best Friend*. New York: Ballantine Books, 2006. Print.

Merleau-Ponty, Maurice. *Phenomenology of Perception*. Trans. Colin Smith. New York: Routledge, 1962. Print.

———. *The Visible and the Invisible*. Trans. Alphonso Lingis. Ed. Claude Lefort. Evanston: Northwestern UP, 1968. Print.

Morgan, Julie. Telephone interview. 3 Dec. 2011.

Myers, Arthur. *Communicating with Animals: The Spiritual Connection between People and Animals*. Chicago: Contemporary Books, 1997. Print.

Ryan, Jeri. Telephone interview. 1 Dec. 2011.

Schoen, Allen. *Kindred Spirits: How the Remarkable Bond between Humans & Animals Can Change the Way We Live*. New York: Broadway Books, 2001. Print.

Schaffer, Michael. *One Nation under Dog.* New York: Henry Holt and Company, 2009. Print.

Sheldrake, Rupert. *Dogs That Know When Their Owners Are Coming Home.* New York: Crown Publishers, 1999. Print.

Smith, Penelope. *When Animals Speak: Techniques for Bonding with Animal Companions.* New York: Atria Paperbacks, 2009. Print.

Smyth, Bryan. "Merleau-Ponty and the Generation of Animals." *PhaenEx* 2.2 (2007) 170–215. Print.

Thomas, Keith. *Man and the Natural World: A History of Modern Sensibility.* New York: Pantheon Books, 1983. Print.

Van Manen, Max. *Researching Lived Experience: Human Science for an Action Sensitive Pedagogy.* New York: State U of New York P. Print.

Williams, Marta. *Beyond Words: Talking with Animals and Nature.* Novato, CA: New World Library, 2005. Print.

———. *Learning Their Language: Intuitive Communication with Animals and Nature.* Novato, CA: New World Library, 2003. Print.

———. Telephone interview. 4 Dec. 2011.

Wilson, Edward O. *Biophilia.* Cambridge, MA: Harvard University Press, 1984. Print.

Yeo, Michael. "Perceiving/Reading the Other: Ethical Dimensions." *Merleau-Ponty, Hermeneutics, and Postmodernism.* Ed. Thomas W. Busch and Shaun Gallagher. New York: State U of New York P, 1992. Print.

12 Thinking through Ravens
Human Hunters, Wolf-Birds and Embodied Communication

Pat Munday

Common Ravens (*Corvus corax*) have been a part of my daily life for twenty years—ever since I moved to the once-great mining city of Butte, Montana, along the Continental Divide in the northern Rocky Mountains.

My home in nearby Walkerville seems to be in the center of a breeding pair's territory, and they often herald the dawn from the tall spruce tree in front of my bedroom window. My mile-and-a-half walk to work is between a rookery and two food sources—the interstate highway (a prime source of roadkill) and the county landfill. Rare is the day that a juvenile conspiracy (i.e., a flock of ravens) and a mating pair or two does not pass overhead. Winter or summer, from my office window and while walking about the campus of my little college, I enjoy the sight and sounds of ravens.

I like ravens. They are smart, inquisitive but cautious, familiar with a large territory and seem to like the same kinds of places where I prefer to live and roam. Cross-country and backcountry skiing, trout fishing, backpacking and hiking, and hunting define the four seasons of my year. While I am engaged in these activities throughout the mountains, prairies and rivers of southwest Montana, ravens are usually part of the picture. One of the largest conspiracies I've seen consisted of seventy or more ravens.[1] They assembled noisily on top of West Goat Peak, the highest mountain in the Anaconda-Pintler Wilderness, just before a large forest fire blew up around noon on a hot mid-August day in the year 2000. My young daughter and I had summited the peak while on a backpacking trip. We—along with the ravens—witnessed a large mushroom cloud tower into the sky twenty-five miles to the west. Charred pine needles rained down from the sky, lifted on the updraft of the fire and carried to us by the prevailing wind. I believe the ravens were as amazed as we were: They looked toward the billowing tower of smoke, hopped about and made short flights while squawking excitedly and investigated the burned, fallen needles.

I like talking with ravens. Though my ravenish vocabulary is limited to a fair quork, a good alarm scream and a passable croak, it's enough to attract passing ravens and occasionally ("What did I say?") to enrage them. A better listener than talker, I've had ravens lead me to various large animal carcasses and—while hunting—to live elk.

"Listen to me," said the Raven, "but it is so difficult to speak your language! Do you understand Ravenish? If so, I can tell you much better." Hans Christian Andersen, like J. R. R. Tolkien and other writers, drew on Norse mythology regarding the communicative powers of ravens.[2] The god Odin carried on his shoulders two ravens, Huginn (Thought) and Muninn (Memory), who made a daily flight over the earth acting as Odin's eyes and ears, keeping him informed about world events (Sturluson 39). Odin, as the chief god in Nordic mythology, dates to at least the sixth century and, along with the similar Germanic form Wodan or Wutan, persisted well into the Christian era (Davidson, "Lost Beliefs" 60, 79). In addition to receiving news of the world from his two ravens, Odin was a shaman who could travel in the form of birds or animals (77).

Like Odin, mortals could also learn much from observing ravens and other birds (Davidson, "Lost Beliefs" 137). Reading and interpreting bird behavior as omens extended well south from Scandinavia to the Greco-Roman world and other premodern European cultures. Ravens and other corvids frequently figured into auguries described by Cicero, Livius and Pliny. As religious officials, augurs read birdcalls and behaviors to foretell future events ("Augur, Augurium" 174–179). Given the raven's connection with death as a scavenger of human corpses, it seemed a fitting vehicle through which to communicate with the gods of the Otherworld and souls of the dead. Ovid's writings emphasized the untrustworthiness of the raven as a messenger bird as well as its being a bearer of bad news (Newlands 244–255). At a practical level, the raven served two primary functions in premodern European life: It heralded the dawn and it scavenged dead and dying soldiers on the battlefield. The author of *Beowulf* as well as Roman writers such as Sidonius Apollinaris and Martialis sometimes began their stories with the raven's morning matins (Hume 60–63; Martin 76).[3]

In modern Western civilization, ravens acquired a decidedly bad rap as an untrustworthy and evil bird. This can be traced to early Christian and Jewish interpretations of the raven's role in helping Noah find land after the flood. Despite the way that later scholars demonized ravens, Genesis 8:7 is ambiguous on this point, stating only that Noah "sent out the raven; it went to and fro until the waters had dried up from the earth" (Torah 14). After this, Noah sent forth a series of three doves with more definitive results. As a precursor to and possible template for the biblical flood story, Utnapishtim in the Babylonian *Epic of Gilgamesh* also sent out a raven and a dove (in addition to a swallow). For Utnapishtim, however, it is the raven that communicates the vital information that the waters have receded (Mitchell 118; Freedman 124, 127–129).

Nonetheless, theologians from the fourth century on generally wrote that the raven had failed Noah in its mission and characterized it as "an unclean bird"; "a symbol of evil"; and "the enemy," representing those destroyed by the flood (Marcus 71–80; Horowitz 504–505). Apparently it was easy to ignore the rather nobler role of ravens in Old Testament literature, as when a raven was commanded by god to feed the prophet Elijah during his sojourn in the wilderness (Torah, 1 Kings 17:4,6). By the seventeenth century, the raven's reputation in Europe was sealed. Though ravens may have played a crucial and positive role in scavenging human carcasses after the London's Great Fire of 1666, the sight seems to have provoked only hatred and revulsion for the bird (Sax 102–105). Bounties and more indirect measures (such as forest clearing) led to extinction throughout much of England and the rest of Europe (Ratcliff 20–24; Marzluff and Angell 10). Though ravens are now protected in much of Europe, many Germans believe they kill too many nestlings and even calves, and are pressuring to have them removed from the endangered species list (Heinrich, *Mind of the Raven* 144–145).

Western science and philosophy severed connections—and communication—between humans and other animals.[4] Descartes, Francis Bacon and other architects of the early modern period facilitated this intellectual dichotomization of nature/culture, as has been well documented by Carolyn Merchant and other historians (Merchant; Oelschlaeger 68–96). In practical terms, as Richard Louv and Paul Shepard have explored, developments such as industrialization, urbanization and consumerism also reshaped society in ways that reinforced this disconnection. Postmodern approaches seem to take this disconnect for granted, as do advocates of social construction who "rely on a humanist perspective about knowledge creation that privileges the cognitive sovereignty of human subject over nature" (Crist 5).[5] As environmental problems loom larger, healing the divide between humans and other animals is an important aspect of addressing our alienation from nature.

In this essay, I seek to rehabilitate the relationship between humans and other animals, using the raven as an example of how we, as a culture, might proceed. As an ethnographic model, we can look to indigenous cultures throughout the circumpolar Taiga and Boreal forest biome. Many of these cultures feature ravens (and their close relative, the Wisakedjak or gray jay) in myths and as important totem animals. Often, Raven is endowed with speech or other powers of communication (Oosten and Laugrand; Schwan).[6] Anthropologist Richard K. Nelson introduces his account of a year spent with the Koyukon people of central Alaska with a meditation on ravens:

> During the many months I had spent among the Koyukon, I had gradually begun to look quite differently at ravens, as I began not only to know about, but also to *feel* the further dimension in nature that was so preeminently important to my teachers. Ravens had become more than just beautiful and intelligent birds. I found myself watching them and feeling watched in return . . .watched by something more than the ravens' gleaming black eyes. I found myself listening to their calls, not just to enjoy their strange ventriloqual gurglings but also to hear what they might be saying. (xiii)

The Koyukon—like the Tlingit, Haida and other northern peoples—take ravens seriously. Raven is a central figure in myth, with roles ranging from bringing light to the world, to creating and then altering the world, to playing the trickster. The elevation of Raven in myth does not obscure a practical understanding of raven-ness. People who deceive others in order to fulfill their own needs and those who are not good hunters may be mocked as being "just like a raven" or even scorned with an imitation of a raven's call (Nelson 80–81). At a practical level, hunters look to the flight of a passing raven as a sign of whether the hunt will be successful. They may call out to the raven to ask for luck. Ethologist Bernd Heinrich quotes an Athapaskan woman: "One of the things we say to raven while we hunt is 'tseek'aal, sits'a nohaaltee'ogh,' which means 'Grandpa, drop a pack to me.' If the bird caws and rolls it is a sign of good luck" (Heinrich, *Ravens in Winter* 25).[7] In rationalizing this, a Koyukon explained, "It's just like talking to God, that's why we talk to the raven. He created the world" (qtd. in Nelson 83).

Heinrich also visited with Inuit peoples at several villages on Baffin Island north of Hudson Bay. Like the Koyukon, these Inuit hunters also described speaking to ravens. Echoing my own experience as a hunter, the Inuit and Koyukon report that ravens will then sometimes dip a wing (half roll) and lead them to prey (Heinrich, *Mind of the Raven* 245–254). The engagement between human hunters and ravens is similar to the relationship between ravens and nonhuman hunters such as wolves.

Following the reintroduction of gray wolves in 1994, Yellowstone National Park proved fecund ground for studying raven-wolf interactions. The abundance of elk made for relatively easy wolf kills, with food so abundant that they often ate only the choicest parts before moving on to new prey. Ravens, along with other scavenger birds such as bald eagles and golden eagles, were quick to capitalize on these carcasses. Typically more than thirty ravens fed at a given kill. Wolf researchers found they could locate wolves, and their fresh kills, by watching for flocks of ravens. Heinrich, who had observed great wariness on the part of New England ravens in approaching carcasses, was amazed to find ravens flew over wolves while they killed an elk and then immediately descended to feed with the wolves. He speculates that the coevolution of ravens and wolves—hence the colloquial name for ravens as "wolf-bird"—makes ravens uncomfortable and wary in a world *without* wolves (Heinrich, *Mind of the Raven* 226–235).

Ravens benefit from their association with wolves primarily through 'kleptoparasitic foraging strategy,' a wonderful term for ravens stealing food from wolves. One or two ravens almost invariably accompany wolves while they are hunting. Once an animal is killed, the raven 'scout' calls out to recruit other ravens to the carcass. Thus, flocks of ravens discover wolf-killed carcasses almost immediately. If ravens find a large mammal carcass, perhaps of an animal that has died of starvation in winter or from other natural causes, it is of no avail since the raven's beak is not strong or sharp enough to tear the hide so that ravens can feed. Hence they are dependent upon wolves or other large predators for a major share of their food, especially in winter when smaller prey is scarce. Though ravens do not seem to follow elk or other large prey, they are "quick to locate and harass injured elk, apparently drawing the attention of wolves and coyotes" (Stahler, Heinrich, and Smith 283–290). Ravens also frequently associate with wolves away from kill sites, engaging in playful behavior such as biting at wolves' tails and games of chase with wolf pups (Stahler). Ethologists interpret this 'play' as a means for ravens to learn how to interact safely with large and potentially deadly predators (Heinrich, *Mind of the Raven* 236–237, 292–294).

Wolves, while tolerant of ravens, certainly do not kill *for* ravens. On the contrary, there is good evidence that wolves have evolved into social, pack-oriented creatures in order to maximize meat consumption, thereby leaving as little as possible for scavengers. While this did not seem important with the artificially high elk populations that wolves found in Yellowstone the first few years after being reintroduced into that ecosystem, prey is usually not so abundant and during certain seasons is particularly scarce. A flock of ravens can consume eighty pounds or more of meat from a carcass in a single day (they cache much of it), and over time that has put considerable selection pressure on wolf evolution. Thus the "reduced per capita rate of consumption in larger packs is offset by the benefit of increased frequency of prey capture and reduced loss of food to scavengers" (Vucetich, Peterson, and Waite 1123). The optimal pack size, which depends upon the amount lost to scavengers as well as the size and abundance of prey, can consist of twelve or more wolves. Based on studies like this, Marzluff and Angell speculate that coevolution with ravens played a similar role in favoring group living for humans (285–286).

Ravens also hunt cooperatively in groups with each other, especially for smaller prey such as squirrels (Heinrich, *Mind of the Raven* 132–133). Heinrich also describes ravens flushing flickers, which are then killed by peregrine falcons, with the ravens feeding on whatever the peregrine leaves (139–140). As with wolves or human hunters, such associations are hardly altruistic on the part of ravens. Heinrich even believes there are cases where ravens have lured predators such as mountain lions and grizzly bears to human prey (*Mind of the Raven* 192–195).

Whether it was humans killing other humans on the battlefields of pre-modern Europe or wolves killing elk in Yellowstone National Park, ravens have learned to associate with human and nonhuman predators in order to obtain food. Prior to the reintroduction of wolves to Yellowstone, elk populations grew artificially high, beyond the carrying capacity of the land to support them. In winter, one elk herd migrated annually from the park to the Jackson Hole, Wyoming area. The artificial feeding grounds created as a tourist attraction attracted them. High elk density near Jackson Hole made for high hunter success and an unusually high concentration of gut piles during hunting season. Ravens migrated to the area during hunting season to take advantage of this food source, roughly tripling their population during that time (White).[8] Furthermore, ravens in other hunting areas outside the park boundary learned a foraging strategy whereby they were attracted by gunshots. As wildlife biologist Crow White titled an article about this phenomenon, "Hunters ring the dinner bell for ravens."[9]

Ravens make a wide range of calls, apparently to communicate with one another. While it is difficult to know what a specific call "means" to other ravens, there is good evidence for how this communication affects other ravens and shapes raven society. Ravens typically breed between the age of three and seven years, mating for life and establishing a territory that they defend from other ravens. Prior to choosing a mate, they form loosely organized juvenile "gangs" (i.e., "conspiracies") of anywhere from a dozen to a hundred or more birds (Heinrich, *Mind of the Raven* 108) These gangs are composed of vagrant individuals that may fly more than sixty miles each day to find food. When one or more of these individuals finds a large food source, such as a big game carcass, they yell in order to recruit other juveniles. By feeding as a crowd, the juveniles overwhelm territorial adults (Heinrich, "Winter Foraging" 141–156; Heinrich and Marzluff).

Despite increasing scholarly attention to raven vocalization, no one has yet published the Human-Ravenish Dictionary. Like that of other birds, raven syntax has been found to vary geographically (Balaban). For example, North American ravens make certain common sounds that have not been documented at all in European populations. The use of recording equipment and sonograms—complete with complex analysis of frequency range, call duration and harmonics—shows great promise as a tool for studying intraspecies vocalization (Conner). Because these recordings have been made available in an online archive, ornithologists and ethologists may someday be able to map and decipher raven language.[10] This may prove an even more difficult enterprise than it seems at first glance, however, as even mated pairs seem to develop their own distinctive vocabulary and syntax (Enggist-Dueblin and Pfister, "Communications").

In engaging with wild ravens, we human animals recognize something like ourselves.[11] Raven intelligence, vocalization, playfulness and socialization make them appear very humanlike, and it is not surprising that cultures from the Anglo-Saxon to the Inuit to the Athabascan all cast ravens in powerful mythological roles. Recent scholarship in animal cognition and the evolution of intelligence supports human recognition of ravens as a kindred animal. In assessing the evidence, Nathan J. Emery even poses the rhetorical question "Are corvids 'feathered apes'?"[12] In considering how animals can recognize meaning in the behavior of other species, the semiotician Jakob von Uexküll paraphrased Goethe in saying:

If the flower were not bee-like
And the bee were not flower-like
The unison could never be successful. (65)

Despite difficulties in studying communication between humans and other animals, environmental communication scholars are uniquely situated with theories and methodologies to help bridge the human-nonhuman animal divide. My particular interest is in revealing perspectives on engagement between wild ravens and human hunters. Additionally, as what David Petersen calls a "spiritual hunter," I draw on my own engagement with wild nature (53–57).[13]

From a personal perspective, I have a long and somewhat checkered history of engagement with ravens and other wild birds. The summer of my fourth birthday, Uncle Jim gave me a popgun that fired corks. I remember lying in ambush in the backyard where Grandma Pansy scattered bread-crumbs for the sparrows. I don't think I ever hit any of them, but I do recall the alarmed look they would flash as I prepared to shoot at them—just before they flew off.

At age ten I was somewhat more successful. Roger Smith, a family friend, gave me a "toy" bow he crafted from a hickory tree sapling. My neighbor Ronnie Knott and I couldn't wait to try it out, but Grandma Beryl forbid me from shooting it around the house or yard where I lived. Ronnie's father, a member of the Seneca Nation, worked with Gramps tending an oil lease in the wooded hills of Cobb Hollow, in New York State just north of our home in Pennsylvania. Ronnie and I often accompanied them at work, where we were set free to run the hills. Elderly Mr. and Mrs. Cobb, the leaseholders, lived in a large stone house on the property but they were away during this particular time. They had a bird feeder near the back door. Gramps often knew what we might do before we did it, though it took some years for me to appreciate this prognostic gift. His parting words were "Don't shoot birds and stay away from the Cobb house." The warning forgotten, Ronnie and I found ourselves drawn to a flock of birds near the house soon after Gramps and Mr. Knott drove the Willy's Jeep up the hollow to tend the oil wells. A chickadee perched on the feeder. I took careful aim, let a blunt arrow fly and knocked the little bird off its perch—killing it. The arrow also sailed through the window of the kitchen door. As we held the tiny bird in our hands, Ronnie and I felt terribly guilty—more for having killed a beautiful creature than for the broken windowpane.

The summer I turned twelve, I was walking Gramps' oil lease on the rich bottomland near Tunungwant Creek, carrying a single-shot, twenty-gauge Winchester shotgun. I had dreams of becoming a crack wingshot and was allowed to kill the European Starlings that roosted in the dead elm trees. Several crows flew over a small clearing and I mounted the shotgun and killed one. For the next several years I could not walk the area—especially if I carried a firearm—without the crows harassing me (while they stayed safely out of shotgun range). This was a huge problem when I began bow hunting for white-tailed deer the following year. Despite the abundance of deer around the lease, I had to hunt a nearby ridge instead. The crow experience taught me a great lesson about the intelligence, communicative powers and vindictiveness of birds—or at least of corvids. I became more careful and considerate of them after this time.

Several years later I was walking the banks of the Tunungwant when I saw a remarkable sight. A bird hovered over the water, flapping just enough to maintain altitude and occasionally change position. It tucked its wings, dove into the creek and emerged with a large fish—a white sucker—in its talons. I was stunned. Running back to the oil-lease powerhouse, I described this strange hawklike bird to Gramps. He could not imagine what it might be and, I think, believed I was exaggerating or somehow deranged. The regulation of organochlorines had only recently begun and DDT—though its use had been greatly reduced—had not yet been entirely banned. But already, raptors—such as the osprey I had seen—were recovering.

Wild turkey populations were also recovering from more than a century of near extinction, and were becoming especially prevalent on the Allegheny Plateau just south of our home. Gramps and I, along with his friend Bernard Dutka, began hunting them during their spring mating season. Mr. Dutka taught me about paying close attention to everything that was going on in the woods. Before dawn, crows would begin calling as they left their roost, and would often mark the location of a turkey flock. By listening for the crows and moving quickly into position, it was possible to set up near the turkeys and begin hen-calling at first light when the gobblers were most susceptible to being fooled into range.

The summer I turned sixteen, I passed the driving test and could go off hunting on my own. Still-hunting deer with a rifle, I often encountered flocks of cheerful, friendly chickadees. Following Mr. Dutka's lead, I wondered what sort of sign this might be. The chickadees flocked to me and then flew off, often returning one or more times. One day I carefully and slowly followed them, and they led me to a bedded deer like ravens leading wolves to elk. This was no act of altruism, for the chickadees were often the first bird to arrive at and feed upon the gut pile left by a human hunter. They especially favored suet-rich fat from the body cavity, and I began the practice of hanging it on a tree so that my helper-chickadees could have it to themselves and not share it with a gluttonous fox.

A few years later, while hunting turkeys during spring break from college, I had an experience that was as perplexing and as mystical as having seen the osprey. I was on my way back to the car, stumbling several miles across the marshy ground of the Allegheny Plateau after an early morning hunt. A crow-like bird flew over at a great altitude and called out with a deeply resonant, knock-like voice. I explained this in detail to my grandfather. As with the osprey, he could not imagine what it might have been. Soon after that time, while deer hunting with me and a college friend in the Adirondacks of New York State, he pointed out several ravens circling a high rocky peak and I realized what I had seen. Like the osprey, ravens were (and still are) recolonizing their former range. As a child of the woods, my initial inability to identify the bird left a great impression. It was like coming upon a new word while reading a book, and it was about this time that I began buying field identification guides to extend my knowledge of and appreciation for the natural world of my experience.

Upon moving to Butte, Montana, more than twenty years ago, I was struck by the prevalence of ravens both in town and in the surrounding mountains and forests. Based on what I had learned from chickadees and crows as a young man in the Alleghenies, I paid attention to ravens while elk hunting. Ravens, along with the Wisakedjaks that frequent the thick northern-facing benches where elk bed, led me to many elk. Several years after moving to Butte, I held a class discussion about the human relationship with nature. The conversation turned from questions about "What is a weed?" to the topic of birds. I explained my reverence for and feeling of connection with birds such as ravens. A middle-aged student named Charlie expressed his contempt for ravens as "filthy birds" and told of shooting them whenever he got the chance. Charlie also was, as it turned out, a singularly unsuccessful elk hunter, having never killed one in decades of hunting. I could not help but think that his hatred for a species like the raven represented a deeper dis-ease with nature—a disease that expresses itself at the individual level as well as at the broader, cultural level of Western civilization as a whole.

The semiosis of ravens as evil or filthy birds is a powerful symptom of the modernist disease. For us postmoderns, semiotic analysis is a key first step to investigate engagement and communication between humans and other animals. The structuralist approach pioneered by Saussure and developed by Lévi-Strauss and others unfolds through the dialectical juxtaposition of opposing signs that give rise to new signs and meanings. This method proved especially fruitful for understanding myth and other cultural narratives.

Lévi-Strauss, in pondering the relationship between humans and other animals, wrote, "natural species are chosen not because they are 'good to eat' but because they are 'good to think'" (*Totemism* 89). As an anthropologist heavily influenced by semiotics, Lévi-Strauss theorized that we "think with animals" as a structurally embodied sign. In other words, in the human mind, nonhuman animals—especially wild animals—serve as a mental construct in a larger *mentalité* or *Innenwelt*, especially in cultural myths. He pioneered this approach by analyzing the way in which "gross constituent units"—meaningful elements such as life, death, food and hunting—form patterns in various versions of stories such as the Oedipus myth or the Zuni emergence myth ("Structural Study of Myth" 428–444).

There are other useful ways to apply semiotics to improve our understanding of the role of ravens in Beowulf and related legendary stories. Odin was a powerful and not necessarily benevolent entity, and so naming either places or persons after this god may have been avoided (Davidson, *Lost Beliefs* 56–59).[14] Instead, authors could use metonymy based on Odin as Raven-god, resulting in the naming of places or people after Raven (i.e., *Hrafn*) instead of Odin *per se*. As in the bard's story, the sign vehicle "Raven" was sufficient to evoke the idea or interpretant "Odin" in the names of persons or places. For a culture steeped in oral traditions about Odin or human warriors such as Beowulf, the metonymic meanings of the raven sign vehicle were rooted in indexical knowledge of raven myth as well as the actual bird's behavior. Although "only" a mental construct, Odin's messenger ravens Huginn and Muninn connected participants in Nordic myth back to Odin as surely as a thunderclap connected the participants back to Thor and his hammer.

In Lévi-Strauss's terms, ravens are "good to think" both because of their similarity to us and because of their practical usefulness in helping us find game (if we are a hunter) or find land (if we are a sailor). But Lévi-Strauss was primarily interested in the role of animals in synchronic myth and that is not the primary focus of this paper. Instead, I am interested in the ways we engage with ravens and construct meaning as an ongoing, diachronic process.

As structurally embodied signs, our patterned experience with ravens (in traditional cultures that necessarily entail a synchronic dimension) forms the basis for an ongoing cognitive relationship. We hear the raven's diverse vocalizations and, based on our own vocabulary and syntax, construct it as a sign of raven intelligence. We observe the raven leading us to our quarry and, based on our own cooperation with other human hunters, construct it as a sign of raven intention and connection. The peculiarly similar association of ravens with wolves, from which humans derived dogs as companion animals, further reinforces this cognitive relationship. We talk with our dogs and use them to find drugs or game, and thus dogs and ravens and wolves can become inextricably linked in the stories we tell to make sense of the world.

Therefore I agree with ecosemiotician Andreas Roepstorff that Lévi-Strauss didn't have it quite right in limiting semiotic analysis to the synchronic dimension of myth, for semiotic analysis can also reveal much about our diachronic interaction with wild nature. Using examples from Greenlanders' pragmatic interaction with animals, Roepstorff argues that—for humans who live their lives enmeshed with wild animals—it can be a matter of thinking *along with* animals. For example, Greenland halibut fishermen watch seals to know when a glacier is becoming dangerously unstable and it is time to move from the area. In this way, the Greenlanders construct meaning through constructive mental interplay with the seals.

To build on Roepstorff, ravens are not merely an object of thought in totemism or some other animal-human relationship that we mentally construct and then enact as a mythic narrative. Instead, we can interact with wild animals such as ravens on a day-to-day basis, construct meaning as we go and try to understand the world through them. While a raven *might* make a classic wing-dip when we call out to it, it might also engage us in a host of other ways not explained by formulaic myths or conventions. I was once midway through my walk home when a small flock of ravens flew over. I called out to them expecting a few calls in reply or perhaps for one or two of them to break formation and circle me before moving off with the flock. Instead, the whole lot of them swooped down, landed in some nearby trees and proceeded to hoot and holler in an agitated, angry manner. "Geez," I thought, "what did I say?" I pondered the exchange for the next mile but there was no simple explanation for what had occurred.

Similarly, I rarely climb a mountain peak without having a flock of ravens show up as I scramble along the ridge. They play on the thermals and sometimes pop up from the opposite side of the ridge, fly very close and seemingly take a perverse delight in surprising me. Raven joy is as apparent as a playful dog's joy: They approach from behind, call out as they fly above the ridgeline and then quickly drop down as I turn to see them. This is repeated. What are they up to? Do they expect me to kill and share a mountain goat with them? Are they 'pulling my tail,' the way they tease wolves? It remains a mystery, but is fun to think about. It *is* constructive interplay.

In considering the possibility of human–wild animal communication, Marietta Radomska begins with a quotation from Ludwig Wittgenstein: "If a lion could talk, we could not understand him" (190). The quotation frames the modernist dilemma of communicating with other species, aliens or even human animals with a language very different from our own. It also challenges postmodern perspectives on the social construction of meaning. Radomska employs Charles Peirce's semiotic triad—sign vehicle, interpretant and object—to explore semiotic contact with nonhuman animals and the process of decoding their signs in ways that transcend mere language. Peirce's semiotic triad is an especially useful tool in this analysis because of the emphasis Peirce put on the model wherein objects are the "occasions and results of active experience. To discover an object means to discover the way in which we operate on the world producing objects or producing practical uses of them" (Eco, "Peirce's Notion" 1465). In my analysis, the human hunter watches the raven dip its wing (i.e., "drop a pack") and takes this performative act to be a sign vehicle. The interpretant is the sense that the elk hunter makes of the sign—in this case, the notion that the raven is pointing out the existence of prey and will, perhaps, even lead the hunter to it. The object is presumably the physical elk that the raven "promises," or perhaps the physical ensemble of human, raven and elk (or other quarry). In this case, the Peircean object is a *real* object or objects directly linked through the indexical mode of the sign given by the raven.[15] As an act of semiosis, the hunter's behavioral response to the sign vehicle is an embodied habit and connects the hunter with other beings in a material and semiotic reality. The hunter reads the raven's wing-dip as surely as he or she would read an elk's track or the performative act of a fellow hunter who pointed and said "elk." As Umberto Eco wrote of this sense of embodiment, "to have understood a sign as a rule through the series of its interpretants, means to have acquired the habit to act according to the prescription given by the sign" (1466). It is, in short, the practice of communication.

When a raven dips its wing, rolling while in flight, it does look like a person rolling their shoulder to drop a pack. Ravens, of course, do not carry packs. Nor is it likely that ravens emulate the actions of humans that do so. Nonetheless, as a sign vehicle the raven's wing-dip is common enough and similar enough to a human act that it has become an iconic sign vehicle, signifying the raven dropping a gift to a human hunter. We can speculate as to what the raven might be thinking.

What sort of story does a raven tell itself (and other ravens) about wolves or human hunters? Though we cannot know this in a direct way, we do have access to how ravens see or read the world, just as we see through the eyes (or noses) of our companion animals. Through their performative acts, trained companion species—such as Seeing Eye® dogs, drug sniffing dogs or fox hounds—commonly provide us another animal's view of the world. These are pragmatic processes with objective outcomes: The guide dog signals to its blind human companion to step up; the K-9 dog at a customs gate signals to a human handler that there are illegal drugs in a suitcase; the foxhound signals to a human hunter that it has found the hot, fresh scent of a fox.[16]

Signs are, of course, everywhere in nature and not limited to what human animals experience and read: Consider the physical track or chemical scent of the fox, or the visual infrared "bull's-eye" that marks a flower for a bee.[17] As Deely writes, "the animal's survival depends upon getting right the manner in which the physical environment is incorporated into its world of objects, its Umwelt, when it comes to food, sex, and danger" (28). In this way, the *Umwelt* is the biological foundation for semiosis. In a larger sense, species evolved means for communicating through signs to other members of their species or to certain other species. This is what Jakob von Uexküll and Thomas Sebeok pioneered as the fields of zoosemiotics, biosemiotics and—most recently—ecosemiotics (Radomska; Sebeok, *Global Semiotics*). Some of the emphasis has been on animal-human communication, beginning with Sebeok's research directive, "there is no doubt that the inner human world [our mental *Innenwelt*], with great effort and serious study, may reach an understanding of non-human worlds and of its connection with them" (qtd. in Petrilli and Ponzio 20).[18] On this belief, Sebeok built a positive, postmodern programme that can help us reconstruct the divide between human and nonhuman animals, and heal the Cartesian subject-object split.

Thinking through ravens can be a first step in healing this split. In applying Peirce's triadic model (vehicle, interpretant, object) to the Uexkülls's theory of *Umwelt* (i.e., the environment surrounding an organism), the sign is composed of three elements, each with its own objective independence (Deely 27–28).[19] For the interaction between the human hunter and the raven, our 'internatural' or biosemiotic analysis is a function of standpoint. From the raven's standpoint, when it quorks and dips a wing, embodied in this act of communication is the triad of sign vehicle, interpretant and object: The human hunter is a sign-vehicle, the future possibility of the hunter killing an elk is an interpretant and the hoped-for gut pile is an object. From the standpoint of the hunter, the raven's quork and wing-dip is a sign vehicle, the unseen but imagined elk (as quarry) is the interpretant and the hoped-for act of killing an elk is the object.

In extending the field of biosemiotics established by his father Jakob, Thure von Uexküll wrote, "the exciting fact [is] that signs are in reality magic formulae whose creative power changes our world and ourselves" (12). It is magical to watch interactions between wolves and ravens or to engage as a human hunter in similar direct interactions with ravens. Semiotics provides a theoretical tool for shifting our view of, and engagement with, ravens. Like Aldo Leopold, who watched a wolf die and learned to "think like a mountain" in listening to the howl of a wolf, material-semiotic engagement with wild nature can profoundly change our relationship to the world. From this new perspective, we can imagine a new future.

Imagining the future is a form of storytelling (Buell). Outdoor magazines are, of course, full of stories where hunters imagine killing an elk or other animal. Sometimes these stories even rise to the level of literature, as with Dinesen's *Out of Africa* or Faulkner's "The Bear." Is the human author's story so unlike the narrative a raven imagines when it sees a wolf or human hunter trudging through the snow? Though we have no direct access to what a raven thinks and imagines, the ability of a raven to communicate the presence of an elk—and even to lead us to it—clearly points to the raven as author of a story. The raven models a possible world and communicates that story to another species. It is a practical story that can result in a full belly (both for the raven and human hunter or wolf). The role of the raven in creating and telling this story, casting a human hunter into it as a character and helping shape its conclusion, is an amazing performative act of human–wild animal engagement.

When foxhounds hunt with their human companions, the hounds have "a kind of awareness of their human-constructed 'foxhoundness'" (Baker 199). The hounds employ natural instincts shaped by generations of purposeful breeding: The hound *knows* what is expected of it (Marvin 288). Do ravens have a similar awareness? Do they *know* what wolves or human hunters expect of them? As wild animals, their long coevolution with wolves seems to have resulted in a sort of semiotic symbiosis, whereby the ravens and wolves understand one another and each plays a role in their mutually shared *Umwelt*. As humans migrated to North America over the past twenty thousand years or so, they entered into an old association between ravens and wolves. Heinrich writes, "As far as Raven was concerned, Man, the new predator, was probably just a surrogate wolf who also usually hunted in packs" (*Mind of the Raven* 243).

From the early modern period onward, Western civilization developed a deep, dialectical structuralism of good versus evil with little room for ambiguous trickster figures or boundary crossers like the raven of North American mythology. In thinking about the animal-human engagement in wild places, Rosemary McGuire offers this succinct analysis:

> On the one hand, wilderness is depicted as a blank and threatening void, which we have a moral obligation to develop. On the other, it appears as a pristine, Edenic space uncorrupted by human influence. The passion each viewpoint has elicited testifies to the dominance of the idea of wilderness in the American consciousness. Yet these viewpoints are so polarized they preclude the understanding of our own role and responsibilities in the natural world (551).

Radiating outward from places like my home in the northern Rocky Mountains, ravens have reinhabited many areas where they were once extinct. From Europe to eastern states in the U.S. such as Ohio and Kentucky, more people are encountering this smart bird and gaining an opportunity to better understand the world through their interactions with it. Many other animals, from the wolves of Yellowstone National Park to the peregrine falcons of the New York City skyline, have been successfully reintroduced into their former range. Whether in our own backyard or in the wilderness, we have an unprecedented opportunity to observe and connect with wild nature. Although we cannot recover (or invent) some lost world of Edgar Rice Burroughs where humans and other animals lived in Edenic harmony, we can draw on communication theory and our innate biophilia to better understand human engagement with other animals in wild environments.

Like our understanding of wilderness or wildness, our understanding of wild creatures needs to develop in subtler, complex and less polarized ways. Ravens are not "filthy, evil birds" and wolves are not "gluttonous, evil animals." Neither are they benevolent gods or morally superior anthropomorphized human animals. They are real, embodied creatures that might just as soon scavenge the carcass of a human or wolf as that of an elk or moose. But through our embodied interactions, we can learn to see and read our complex relationship with wild animals such as ravens. Ravens offer us a way to engage nature as what Haraway calls a material-semiotic natureculture (*Companion Species Manifesto* 20). No wonder that in premodern cultures—both Western and non-Western—ravens played ambiguous roles. Like our fellow humans, ravens can and do deceive us, and we can and do misread them. This is fitting given Umberto Eco's definition of semiotics as "the discipline studying everything which can be used in order to lie" (*Theory of Semiotics* 7). Still, the deceptive or ambiguous nature of ravens does not lessen the value of taking seriously our environmental engagement with them. But though the wolf-bird has been much maligned and inscribed by modernism, we also should not be too quick to accept the inscriptions of premodern Scandinavian cultures or indigenous circumpolar cultures. We can, however, use these inscriptions as a guide to reinhabiting the world in the postmodern era. We need to do more than merely *think* with or *think* through ravens. We need to *live* with ravens.[20]

Thinking with, thinking through and living with ravens help us to grasp our dynamic and complex relationship with nonhuman animals. Building a relationship through communication in this way helps us transcend biocentrism—valuing other living creatures—with ecocentrism, whereby we value the interconnectedness of natural systems.[21] Ravens evolved in a landscape complete with large predators such as wolves, grizzlies and other species of bears, and diverse other species of now extinct megafauna. As John Muir taught us, "When we try to pick out anything by itself, we find it hitched to everything else in the Universe" (110). Raven, or wolf-bird, is hitched to those large predators in a food web that includes their ungulate prey as well as the grasses or woody browse upon which ungulates depend. We human animals are also a part of that interconnected natural system, playing various roles including biological consumer, embedded observer and sometimes even predator or prey.

Ultimately, in thinking with, thinking through and living with ravens, we also need to think about the kinds of stories we tell. As Marzluff and Angell point out, we have a human bond with ravens that developed through a process of cultural coevolution, a bond that lends itself to storytelling. But not all bonds, and hence not all stories, are equal. It is one thing to create a narrative about our interconnectedness with ravens through the garbage we send to the local landfill and the birds' dependence upon that food source.[22] It is another thing to create a narrative about our passive observation of wolves in Yellowstone National Park and how ravens lead them to and benefit from their prey. And it is yet a third thing to create a narrative about our interconnectedness with ravens through the active process of subsistence hunting and the way the birds lead us—as predatory human animals—to, and benefit from, our prey.

NOTES

1. In neighboring cities such as Missoula and Bozeman, Raven's *Corvidae* cousin the Black-billed Magpie (*Pica hudsonia*) more commonly fills this scavenger niche, whereas ravens are relatively scarce
2. The etymology of terms used for a flock of ravens—conspiracy or unkindness—reflects the bird's behavior and mythological reputation (Moore).
3. The ornithologist John James Audubon shared the Danish novelist's fascination for the speech of the Common Raven, *Corvus corax*, in his description of the bird: "What that I could describe to you, reader, the many musical inflections by means of which they hold converse during these amatory excursions!" (Audubon 2, 2–3)
4. It is noteworthy that Martialis was a Spanish-born hunter who probably experienced ravens more directly than many other Roman writers who knew ravens primarily as a literary or cultural construct.
5. On the phrase "humans and other animals," see the Haraway interview in Schneider (119).

6. As a prime example of Eileen Crist's criticism, see William Cronon (ed.). For a more favorable view of how a social constructivist approach might heal the nature/culture divide, see Eder.

7. On the role of Wisakedjak (Gray Jay) for the Cree and other indigenous peoples, see Robert Brightman and D. S. Davidson.

8. Heinrich recounts a similar experience described to him by Craig Comstock, a raven-watcher from Starks, Maine, in *Mind of the Raven* (239). A wildlife scientist who studies animal behavior, or ethology, Bernd Heinrich attracted the attention of both professional biologists and naturalists with his thick description of raven behavior, *Ravens in Winter*, published in 1989. In a later book, Heinrich described Inuit accounts of ravens leading polar bears to dead seals (*Mind of the Raven* 251).

9. The increasing number of wolves in the Jackson Hole area seems to be reducing the elk population and may shift raven associations away from human hunters to wolves. This is very similar to Thomas Sebeok's observation on the importance of proper decoding of indexical signs by an animal for its survival ("Give Me Another Horse" 282).

10. Heinrich writes, "Maine game wardens find ravens the best assistants in helping apprehend poachers" because of the way they immediately flock to and circle a kill (*Mind of the Raven* 242).

11. For example, Cornell University's Ornithology Lab/MacCaulay Library has 168 audio and video recordings of the Common Raven, *Corvus corax*, available in its online collection.

12. In this essay, I have focused on *wild* ravens and avoided the abundant literature on tame birds. Like dogs or domesticated wolves, tame ravens take on (in Haraway's terms) a hybrid natureculture.

13. Cf. Emery and Clayton.

14. The term "spiritual hunter" derives from Stephen R. Kellert's research.

15. The question of place names is complicated by the advent of Christianity, which also played a historical role in the demise of place names alluding to pagan gods.

16. Note that in Saussure's formulation of semiotics, there is only the signifier (i.e., sign vehicle) and signified (i.e., the mental image or "meaning" of the sign). See Chandler (14–22).

17. For useful perspectives on the ways humans and dogs (as our companion species) communicate, see Haraway.

18. Note that, technically, objective things that can be seen or otherwise sensed are sign vehicles, and not signs per se. Following Peirce and Deely, the sign itself is constructed through relationships in the vehicle-object-interpretant triad.

19. In a more narrow sense, language—as linguistic communication—evolved in humans as a powerful modeling system and a highly specialized *Innenwelt* (Deely).

20. Note that Deely disagreed with T. Uexküll, who followed in the Kantian tradition of his father and vehemently rejected Deely's insistence on *Umwelt* as an objective world (Deely 22–23).

21. See Haraway's development of Merleau-Ponty's notion of "infoldings of the flesh" (*When Species Meet* 249ff.).

22. On the biocentrism versus. ecocentrism distinction and the way it framed Earth First!'s focus on saving old growth forests, see DeLuca (55–60).

23. About ravens scavenging from garbage around the village, a Koyukon woman said, "They are like orphans . . . living as helpless tramps in a place where they do not belong, seeming to care little about their self-respect" (Nelson 30).

REFERENCES

Audubon, John James. *Ornithological Biography*. Vol. 2. Edinburgh: Neill and Company, 1834. Print.

"Augur, Augurium." *A Dictionary of Greek and Roman Antiquities*. Ed. William Smith. London: John Murray, 1875: 174–179. Print.

Baker, Steve. "Guest Editor's Introduction: Animals, Representation, and Reality." *Society & Animals* 9 (2001): 189–201. *Animals and Society*. Web. 12 Nov. 2009.

Balaban, Evan. "Bird Song Syntax: Learned Intraspecific Variation is Meaningful." *Proceedings of the National Academy of Science USA* 85 (1988): 3657–3660. Web. 6 May 2010.

Brandenburg, J. *Brother Wolf: A Forgotten Promise*. Minocqua, WI: North World Press, 1993. Print.

Brightman, Robert. "Tricksters and Ethnopoetics." *International Journal of American Linguistics* 55 (1989): 179–203. *JSTOR*. Web. 6 May 2010.

Buell, Lawrence. *The Future of Environmental Criticism: Environmental Crisis and Literary Imagination*. Malden, MA: Wiley-Blackwell, 2005. Print.

Chandler, Daniel. *Semiotics: The Basics*. 2nd edition. New York: Routledge, 2007. Print.

Conner, Richard N. "Vocalizations of Common Ravens in Virginia." *The Condor* 87 (1985): 379–388. Web. 7 May 2010.

Cornell University's Ornithology Lab/MacCaulay Library, online collection of bird vocalizations. N. pag. *MacCaulay Library*. Web. 2 Dec. 2009.

Crist, Eileen. "Against the Social Construction of Nature and Wilderness." *Environmental Ethics* 26 (Spring 2004): 5–24. Print.

Cronon, William. "The Trouble with Wilderness; or, Getting Back to the Wrong Nature." *Uncommon Ground: Rethinking the Human Place in Nature*. Ed. William Cronon. New York: W. W. Norton and Company, 1995: 69–90. Print.

Davidson, D. S. "Some Têtes de Boule Tales." *Journal of American Folklore* 41 (1928): 262–274. *JSTOR*. Web. 24 Nov. 2009.

Davidson, Hilda Ellis. *The Lost Beliefs of Northern Europe*. New York: Routledge, 1993. Print.

Deely, John. "Semiotics and Jakob von Uexküll's Concept of *Umwelt*." *Sign Systems Studies* 32.1/2 (2004): 11–34. Web. 19 May 2010.

DeLuca, Kevin Michael. *Image Politics: The New Rhetoric of Environmental Activism*. New York: Guilford Press, 1999: 55–60. Print.

Dinesen, Isak. *Out of Africa*. New York: Random House, 1937. Print.

Eco, Umberto. "Peirce's Notion of Interpretant." *Comparative Literature* 91 (1976): 1457–1472. *JSTOR*. Web. 22 Oct. 2004.

———. *A Theory of Semiotics*. Bloomington: Indiana UP, 1978. Print.

Eder, Klaus. *The Social Construction of Nature: A Sociology of Ecological Enlightenment*. Trans. Mark Ritter. Thousand Oaks, CA: Sage Publications, Inc., 1996. Print.

Emery, Nathan J. "Are Corvids 'Feathered Apes'?—Cognitive Evolution in Crows, Jays, Rooks and Jackdaws." *Comparative Analysis of Mind*. Ed. Shigeru Watanabe. Tokyo: Keio Press, 2004: 181–213. Web. 24 May 2010.

Emery, Nathan J., and Nicola S. Clayton. "The Mentality of Crow: Convergent Evolution of Intelligence in Corvids and Apes." *Science* 306 (2004): 1903–1907. Web. 25 Nov. 2009.

Enggist-Dueblin, Peter, and Ueli Pfister. "Communication in Ravens (*Corvus corax*): Call Use in Interactions between Pair Partners." *Advances in Ethology: Supplements to Ethology* 32 (1997): 122. Ed. Michael Taborsky and Barbara Taborsky. Print.

Freedman, R. David. "The Dispatch of Reconnaissance Birds in Gilgamesh XI." *Janes* 5 (1973): 123–129.

Haraway, Donna. *The Companion Species Manifesto: Dogs, People, and Significant Otherness.* Chicago: Prickly Paradigm Press, 2003. Print.

———. *When Species Meet.* Minneapolis: U of Minnesota P, 2008. Print.

Heinrich, Bernd. *Mind of the Raven.* New York: HarperCollins, 1999. Print.

———. *Ravens in Winter.* New York: Vintage, 1991. Print.

———. "Winter Foraging at Carcasses by Three Sympatric Corvids, with Emphasis on Recruitment by the Raven, *Corvus corax.*" *Behavioral Ecology and Sociology* 23 (1988): 141–156. *JSTOR.* Web. 02 Dec. 2009.

Heinrich, Bernd, and J. M. Marzluff. "Do Common Ravens Yell Because They Want to Attract Others?" *Behavioral Ecology and Sociology* 28 (1991): 13–21. *JSTOR.* Web. 02 Dec. 2009.

Horowitz, Sylvia Huntley. "The Ravens in *Beowulf.*" *The Journal of English and Germanic Philology* 80 (1981): 502–511. *JSTOR.* Web. 22 Apr. 2010.

Hume, Kathryn. "The Function of the Hrefn Blaca: *Beowulf* 1801." *Modern Philology* 67 (Aug. 1969): 60–63. *JSTOR.* Web. 11 Dec. 2009.

Lévi-Strauss, Claude. "The Structural Study of Myth." *Journal of American Folklore* 68 (1955): 428–444. *JSTOR.* Web. 2 Apr. 2010.

———. *Totemism.* Boston, MA: Beacon Press, 1963. Print.

Louv, Richard. *Last Child in the Woods: Saving Our Children from Nature-Deficit Disorder.* Chapel Hill, NC: Algonquin Books, 2005. Print.

Marcus, David. "The Mission of The Raven (Gen. 8:7)." *Janes* 29 (2002): 71–80. Web. 23 Apr. 2010.

Martin, Earnest Whitney. *The Birds of the Latin Poets.* Palo Alto, CA: Stanford UP, 1914. Print.

Marvin, Garry. "Cultured Killers: Creating and Representing Foxhounds." *Society & Animals* 9 (2001): 273–292. *Animals and Society.* Web. 25 Nov. 2009.

Marzluff, John M., and Tony Angell. "Cultural Coevolution: How the Human Bond with Crows and Ravens Extends Theory and Raises New Questions." *Journal of Ecological Anthropology* 9 (2005): 69–75. Web. 5 Dec. 2011.

———. *In the Company of Crows and Ravens.* New Haven, CT: Yale UP, 2005. Print.

McGuire, Rosemary. "Understanding Wilderness: Humans and Ecology in Alaskan Nature Writing." *Interdisciplinary Studies in Literature and Environment* 16 (2009): 551–567. Web. 23 Nov. 2009.

Merchant, Carolyn. *The Death of Nature: Women, Ecology and the Scientific Revolution.* San Francisco, CA: Harper & Row, 1980. Print.

Mitchell, Stephen, trans. *Gilgamesh: A New English Version.* New York: Free Press, 2004. Print.

Moore, P. G. "Ravens (*Corvus corax corax* L.) in the British Landscape: A Thousand Years of Ecological Biogeography in Place-Names." *Journal of Biogeography* 29 (2002): 1039–1054. Web. 7 May 2010.

Muir, John. *My First Summer in the Sierra.* 1911. San Francisco, CA: Sierra Club Books, 1988. Print.

Nelson, Richard K. *Make Prayers to the Raven: A Koyukon View of the Northern Forest.* Chicago: U of Chicago P, 1983. Print.

Newlands, Carole E. "Ovid's Ravenous Raven." *The Classical Journal* 86 (Feb.–Mar. 1991): 244–255. *JSTOR.* Web. 4 Feb. 2010.

Oelschlaeger, Max. *The Idea of Wilderness: From Prehistory to the Age of Ecology.* New Haven, CT: Yale UP, 1991. Print.

Oosten, Jarich, and Frédéric Laugrand. "The Bringer of Light: The Raven in Inuit Tradition." *Polar Record* 42.3 (2006): 187–204. Web. 11 Dec. 2009.

Petersen, David. *Heartsblood: Hunting, Spirituality, and Wildness in America.* Washington, D.C.: Island Press, 2000. Print.

Petrilli, Susan, and Augusto Ponzio. *Thomas Sebeok and the Signs of Life.* Cambridge, UK: Icon Books, 2001. Print.

Radomska, Marietta. "Zoosemiotics as a New Perspective." *Homo Communicativus* 1 (2006): 71–77. Web. 20 Dec. 2009.

Ratcliffe, Derek. *The Raven: A Natural History in Britain and Ireland.* London: T. & A. D. Poyser, 1997. Print.

Roepstorrf, Andreas. "Thinking with Animals." *Sign Systems Studies* 29.1 (2001): 203–215. Ed. Peter Torop, Milhail Latman, and Kalevi Kull. Web. 24 Nov. 2009.

Sax, Boria. *Crow.* London: Reaktion Books, 2003. Print.

Schneider, Joseph. *Donna Haraway: Live Theory.* New York: Continuum Publishing Company, 2005. Print.

Schwan, Mark. "Raven: The Northern Bird of Paradox." Alaska Department of Fish and Game. Web. 11 Dec. 2009.

Sebeok, Thomas A. "Give Me Another Horse." *Reading Eco: An Anthology.* Ed. R. Capozzi. Bloomington: Indiana UP, 1997: 276–282. Print.

———. *Global Semiotics.* Bloomington: Indiana UP, 2001. Print.

Shepard, Paul. *Nature and Madness.* Athens: U of Georgia P, 1982. Print.

Stahler, Daniel, Bernd Heinrich, and Douglas Smith, "Common Ravens, *Corvus corax*, Preferentially Associate with Grey Wolves, *Canis lupus*, as a Foraging Strategy in Winter." *Animal Behaviour* 64 (2002): 283–290. Print.

Stahler, Daniel. "Interspecific Interactions between the Common Raven (*Corvus corax*) and the Gray Wolf (*Canis lupus*) in Yellowstone National Park, Wyoming: Investigations of a Predator and Scavenger Relationship." MS thesis. University of Vermont, 2000. Print.

Sturluson, Snorri. *The Prose Edda.* Trans. Jesse L. Brock. New York: Penguin, 2006. Print.

The Torah: A New Translation of the Holy Scriptures According to the Masoretic Text. Philadelphia: The Jewish Publication Society of America, 1962. Print.

Uexküll, Jakob von. "The Theory of Meaning." *Semiotica* 42 (1982): 25–82. Print.

Uexküll, Thure von. "Introduction: Meaning and Science in Jakob von Uexküll's Concept of Biology," *Semiotica* 42 (1982): 1–24. Print.

Vucetich, John A., Rolf O. Peterson, and Thomas A. Waite. "Raven Scavenging Favours Group Foraging in Wolves." *Animal Behaviour* 67 (2004): 1117–1126. Print.

White, Crow. "Hunters Ring the Dinner Bell for Ravens: Experimental Evidence of a Unique Foraging Strategy." *Ecology* 86 (2005): 1057–1060. Print.

White, Crow. "Indirect Effects of Elk Harvesting on Ravens in Jackson Hole, Wyoming." *The Journal of Wildlife Management* 70 (2006): 539–545. Print.

Wittgenstein, Ludwig. *Philosophical Investigations.* Trans. G.E.M. Anscombe. Ames, IA: Blackwell Publishing, 1953. Print.

13 Un-Defining Man
The Case for Symbolic Animal Communication

Stephen J. Lind

I am offering my Definition of Man in the hope of either persuading the reader that it fits the bill, or of prompting him to decide what should be added, or subtracted, or in some way modified.

–Kenneth Burke

There are many stories I could tell about my interactions with animals—from the childhood time I spent sitting on the porch, sharing blue raspberry suckers with my cat Furball, to the puppy I brought home one early Saturday morning on my bicycle (I had been asking my parents if I could have one—finally my still-sleeping father muttered, "Just go get it. Your mother won't be able to say no once she sees it"). As odd as it may sound to some traditionalists, I would probably say that my relationship with my later cat Rascal was akin to friendship. He was a great indoor-outdoor cat, keeping the critters out of my dad's workshop during the day, then ready to go inside when I would get home from a long day of high school. He was with us for six years, but while I was away at college, some owls moved in to the woods and we never saw him again. I even developed what will likely be a lifelong fascination with squirrels after seeing a street vendor selling chipmunks out of a birdcage on the streets of Paris ('if it weren't for customs regulations . . .'). Many individuals besides me, of course, have had strong relationships with animals, and many more find simply observing their behavior fascinating. I saw the limitations of that truth, however, tested as I went off to the halls of higher education.

In college, I eventually found my way to the communication studies field (I had a difficult time deciding on a major, so I figured I would choose one that would require classes with practical benefits for whatever specific career I ultimately chose). I quickly felt at home in communication studies and have been continually enriched by the discipline. As an undergraduate student,[1] however, looking at journals, popular books and seminal texts, I began to see that there were things within the field that were acceptable to study and things that fell outside of what was considered appropriate scholarship. For as broad, diverse and pervasive as the communication field was, some particular interests were set aside for "others" to study, despite their overlap with the field's interests. This was poignantly true of animal communication. As a young scholar looking at the work in the field, it was clear to me that another rhetorical study of Lincoln's "Gettysburg Address" would be right on track, but a rhetorical analysis of squirrel chatter? Certainly not. The question was never even about the *ability* to study animal communication, but rather the *legitimacy* of the inquiry. At best, it might be feasible and marginally appropriate to look at animal communication as an analogy for human interactions, but to take animal or animal-human communication seriously on its own merit was not, and still *is* not, what one does in the communication field.

Communication as a discipline routinely excludes animal studies from its purview of investigation. The point that I hope to demonstrate here is that such segmentation is shortsighted and is based on flawed conceptual understandings of human and animal communication. What is needed is not so much a *change* of perspective, but an *opening up* of perspective. The field does not need a course correction as much as it needs a *coarse* correction—one that roughs up the surface a bit and opens up new possibilities without defining some new polished and tidy way of operating. This chapter, then, is intended to serve as an intervention. Through a discussion of Kenneth Burke's theoretical views as they are reiterated throughout the communication discipline and an analysis of animal interactions as symbolic communication, this chapter will hopefully serve to broaden the horizons of inquiry, bringing in new, exciting investigations into animal and animal-human communication.

BURKE: THEORY IN BRIEF

Kenneth Burke is one of the most influential twentieth-century communication theorists, with vast contributions across disciplines and subjects. In his "Definition of Man,"[2] Burke argues that symbolic communication is a distinct and uniquely human attribute, one that sets humans apart from other animals. His definition is comprised of five clauses:

1. [Man][3] is the symbol-using animal.
2. Man is the inventor of the negative.
3. Man is separated from his natural condition by instruments of his own making.
4. Man is goaded by the spirit of hierarchy.
5. Man is rotten with perfection.

This definition has retained serious traction in disciplinary scholarship and the textbooks that instruct successive generations of researchers. In *Communication Theories for Everyday Life*, Baldwin, Perry and Moffitt introduce Burke by saying, "His early work as a literary and music critic for the magazine the *Dial* quickly led his precocious mind from the appeal of art and literature to a lifelong exploration of human beings as 'symbol using animals'" (93). Em Griffin says of Burke in *A First Look at Communication Theory* that he "started out by acknowledging our animal nature, but . . .he emphasized this uniquely human ability to create, use, and abuse language" (333). According to Hanno Hardt, writing in *Critical Communication Studies*, "Burke offered views of the individual as a symbol-using animal and insisted throughout his work on the values of knowing the symbol systems that surround and affect the lives of people" (126). Stephen Littlejohn and Karen Foss, in *Theories of Human Communication*, say Burke is "no doubt a giant among symbol theorists" who "views the individual as a biological and neurological being, distinguished by symbol-using behavior" (154). These texts are but a few among the many communication studies texts that habitually invoke Burke's definition.

What is more important than the oft-repeated status of Burke's definition is that most of these works offer no rebuttal to the definition. Instead, it is typically and implicitly accepted as the industry standard. That is not to say that it is not explored, but rather that it routinely goes unchallenged—an important distinction. Recaps and explorations also tend to highlight only the first clause of the definition—that *humans are the symbol-using animals*. Although Burke offers his definition up for criticism, addition and refutation, it has instead avoided such in most instances. This first clause, the distilled and most-remembered portion of the definition, has served to bracket all nonhuman animals from a healthy, inquisitive and open field of study. By defining humans as *the* symbol-using animals, Burke's definition and its subsequent reiterations have locked out a whole realm of exploration into animal and animal-human interaction from the communication studies discipline. This definition (by which I technically mean the first clause alone, given the definition's simplified reiterations) is comprised of two key components: *symbol* and *using*. Burke sets up two categories, *humans* and *animals*, and sets them apart from one another by saying that one is the exclusive user of symbolicity. Human action, in such a view, operates through a relationship of things and ideas—needing symbols to referentially convey meaning instead of telepathically transmitting the essence of what is being expressed (see also Henderson; Murray).

What is meant by this means of mediating reference—the symbol—then, seems important. It is a concept that may at times be taken for granted, often simplified as meaning roughly *anything that referentially points to something beyond itself.* That basic definition has its merits, but in her lecture "How Should We Study the Symbolizing Animal?" Celeste Condit makes the term clearer by contrasting it with the more basic term *sign*:

> Symbols are at least quantitatively different from signs, because signs posit a more direct relationship between the sign and the environmental stimulus to which it is linked (e.g. "where there is smoke, there is fire"). Symbols, however, are interpretable in relationship to the other symbols by which they are contextualized. (12)

Likewise for Burke, symbols are not mere automatized utterance or sign, but rather "might be called a word *invented* by the artist to specify a particular grouping or pattern or emphasizing of experiences" (*Counter-Statement* 153, italics added). The symbol is borne out of meaning-governing context by the choices of the artist, and as Ferdinand de Saussure described, is imbued with motivation (Saussure 64–68). Signs are natural and compulsorily generated, whereas symbols are contextually invented and purposefully used. The difference between signs and symbols is then inherently linked to Burke's conceptual and terminological distinctions between *motion* and *action*. Burke explains that "the basic unit of action would be defined as 'the human body in *conscious* or *purposive* motion'" (*Grammar of Motives* 14, italics added), and that "often it is the *preparation* for an act, which would make it a kind of symbolic act" (20). Humans, for Burke, act consciously,[4] not just displaying the natural motions of rolling stones, smoking kindling or schools of fish. In distinction from all other forms of life, humans act. At the heart of this action is the production of communicative items that refer (at least somewhat) abstractly to other concepts, beyond just an instinctual/unmotivated instance of basic cause-effect signing.

The point is not to chart Burke's nuanced conceptions of symbolicity (e.g., the symbolic act as the "dancing of an attitude"—*Philosophy of Literary Form* 9), although he certainly holds to this human/nonhuman :: action/motion split throughout his works,[5] making it clear that for Burke *symbol* and *using* are terms inextricably linked. Instead, the goal here is to highlight the popular and rudimentary definition of humans that has pervaded the communication community in order to see if its core components—symbol and using—might not also apply to nonhuman animal communication. Burke's approach to defining humans is one that has undercut aspects of the field's growth and creativity. It is an approach that I hope to destabilize.

One way of unsettling a flattened notion of animal communication as automatic motion is to simply employ the language of other traditional communication theory. Although Burke's definition is routinely offered as the explanation of the difference between humans and animals, and although Burke's larger work offers immense value to the communication discipline, it may be the case that other terminologies are better suited for the case at hand. If a Burkean understanding of communication has not afforded a perception that animal communication is rich with analytical potential, perhaps another classic model of how humans *do things* will prove useful to understanding what nonhumans *do*. One might, for instance, use J. L. Austin's speech act theory as a starting point from which to understand communication in a way that is both traditional and open-minded. Austin's speech act theory provides a view of communication that is in many ways harmonious with Burke's, yet it may be more useful for analyses that seek to demonstrate and explore the communicative action of animals.

Like Burke, Austin explains that the use of symbols can be seen as action. In *How to Do Things with Words*, Austin claims that when an individual speaks, he or she is not only symbolically representing something, but rather the utterance itself is often a performative action that has power (6). For instance, in a wedding ceremony, the phrase "I do" contains a contractual and promissory essence that is established through the utterance. The phrase is not merely simple referentiality/representation, but rather it is an act. Austin divides the categories in which statements fall into three possibilities—*locutionary*, *illocutionary* and *perlocutionary* acts (94–110). Locutionary acts are the acts of uttering words that one knows to have meaning. This act merely contains the rhetic act, which is the act of sense and reference. The illocutionary act is the doing of something by saying something, such as the promissory "I do." It is the "performance of an act *in* saying something as opposed to performance of an act *of* saying something" (99–100). Perlocutionary acts are instances when individuals not only say something, but by design say something that brings about specific effects on the hearer. The effects may be on belief, attitude or behavior. An individual may persuade another, for instance, to take a different course of action than he or she would have otherwise taken had there not been the speech act. Although the categories are distinct, any given speech act has the potential to fulfill one or more of them, and the usefulness of these terms will be borne out in the next section of this chapter.

Nuanced terminologies and open perspectives allow for a rich exploration of a subject, as Burke's work typically affords and as Burke himself argued (see *Language as Symbolic Action* 44–62; *Permanence and Change* 69–163). Quick, referential nods to Burke's definition in an article or textbook, however, simply tell burgeoning academics that humans are the animals that use symbols. They are *the* animals that use symbols. The only ones. If someone wants to study other animals (the non-symbol-using ones), he or she should find his or her way to other disciplines in the zoological sciences. The study of symbolic communication, through theories like Burke's or Austin's, it is alleged, belongs to the humanities. The *human*-ities. What I seek to demonstrate here is that this schism is largely artificial. This exclusionary definition itself is "rotten" and has perpetuated flawed divides between studying human communication and animal communication, let alone the exciting possibilities that exist in the overlap. To challenge this gap, I want to offer some ethological examples that cut against the exclusivity of the two central tenants to the discipline's simplified definition of humans—*symbolicity* and *usage*.

SYMBOLICITY: WHEN A GRUNT IS MORE THAN A GRUNT

The concept that animals communicate is actually one to which most people will adhere. Wolves howl at the pack, rattlesnakes jitter their tails and birds sing to one another. That is not the issue at hand. The *type* of communication that they participate in is. It may seem compelling to simply say that all communication is inherently referentially symbolic, and thus animals necessarily communicate symbolically. That may be a little tautological for some; it might also seem reasonable in that our communication remains perpetually mediated absent telepathy and thus always referential. Given Condit's distinction between signs and symbols, however, we should probably seek a more robust set of demonstrative evidence. It is understandable that one might see the rattlesnake's tail as *significant* instead of *symbolic*, in the same way that smoke signifies fire in a very direct manner. Issues of purposiveness obviously fold back onto this discussion (the issue then being whether a rattlesnake has any more motive in producing its rattle than a fire does to produce smoke), but I will hold that line of argumentation off for the time being and instead unpack the symbolism in several examples of animal communication.

Burke attempted to demonstrate the nonsymbolicity of animals by telling a brief anecdote about a bird he had seen. This bird was trapped inside a room, and Burke found himself unable to educate the bird on how to leave through an open window (*Language as Symbolic Action* 3–5). Whereas Burke's investigation of animal-human communication is actually what we are after here, his explanation is at best a straw-person (straw-bird?) argument. Although Burke could not engage the bird in the way *he* wanted to, birds communicate extensively with one another in nuanced and well-documented ways. Studies of bird calls are one of the most classic genres of animal communication studies. In 1968, William Evans took a close look at the calls starlings would make when they were in danger or distressed. According to Evans, the specific calls he observed, distinct from other starling calls and songs, would consistently result in the flock's departure. When a predator or other danger confronted the flock from overhead, a particular *seeee* call would be made and a *chuck* sound would be produced when a threat was present on the ground (80–81). The calls had a distinct perlocutionary effect on the receivers based on the specific content of the vocalization.

Male Hooded Warblers also use audible calls with what one might call specific *referential meaning*. According to a 1998 study by Donald Owings, male Hooded Warblers will produce *chipps* and *weeta weeta weTEEoo* sounds when in the nesting/breeding phase, as competition for breeding territory is continued between males. When a female enters, males can then be heard mixing particular songs into their own loops. If an intruder comes into the nesting area near the young, parents will be heard making high-pitched *chink* sounds (4–6). The specificity of these calls seems to indicate a persistent locutionary aspect to the communication. The sounds are related to very specific situations with seemingly specific references, even specific perlocutionary goals. If one were to describe the same situation, replacing the *chips* with *Howdy!*s and *male Hooded Warbler* with *middle-class human males*, there would be no doubt about the symbolic nature of these calls.

Despite the middle-class comparison, one might still say that the specificity and dependability of these sounds renders them signs, not symbols.

Without yet stepping into the territory of conscious use, allow me to offer a couple more examples of bird songs that seem to hint more at the idiosyncratic symbolic, instead of the generic significant. In a 2008 study of male Grasshopper Sparrows, Proppe and Ritchison concluded that the group they were studying not only had specific types of songs linked to specific behavior, but that this particular type of song is not common amongst other species (4–5). Black-Capped Chickadees have also been recorded having very unique songs. According to Gammon, Hendrick and Baker's 2008 study of songbird repertoires, specific chickadee calls were found in singular locations, with "novel song types frequently appear[ing] in small isolated populations" as a "result of imperfect song-learning combined with geographic isolation" (1003). The point here is that arbitrary and unique calls are developed, learned and understood within a particular group. The meanings have been found to be contextual and colloquial, instead of generic and natural.

If one takes smoke/fire as the emblematic case for *signs*, then one should be able to definitionally exclude these bird songs from that category. In the smoke/fire exemplar, the smoke is an effect of the fire. It is not a referential locutionary utterance engineered for perlocutionary effect. Even still holding off on the consciousness issue, the ethological studies of bird songs attest to an undeniably specific purpose for individual songs. In fact, these songs are not even universal across different groups of the same species of birds. Instead, they are unique, interpretable and referential cues that must be understood within their particular context. For definitional intents and purposes, one might even want to call them symbols.

But, lest I rest my ethological laurels on the symbolic songs of birds alone, a few more types of animal communication will be explored here to demonstrate the breadth of symbolicity across the animal kingdom. Marine mammals, for instance, are often cited for their unique communication abilities. Underwater transmissions are integral to these animals. As Evans notes, "the entire life of the whale is intimately bound up with [its] ability to perceive and produce sounds" (126). Gray whales, for instance, recognize and flee at a particular sound emitted by orcas (Davis 142). Similarly, Rossong and Terhune have observed the importance of calls for harp seals. These calls, often transmitted during the breeding season, serve to form the seal herd and can be detected by seals at extremely long ranges, tens of kilometers in diameter (615–616). The vocalizations of these animals play important roles when members of the community appropriately understand their meaning.

In a bizarre 1960s attempt to create/simulate human-animal communication, Dwight W. Batteau created an electronic device that took human speech and transmuted it into whistle patterns. Various formulaic instructions were then contrived using a vocabulary of whistled words created by Batteau speaking into the machine. Commands such as "hit-the-ball-with-flipper" were coded into clicks and whistles of Batteau's choosing and were taught to the dolphins (Davis 148). It should certainly be noted that Batteau did train the dolphins via a reward method, denying any notion of a breakthrough in cross-species translation. In his own backward way, however, Batteau did point to at least the *capacity* for these animals to recognize arbitrary noises as contextually meaningful, perlocutionally referencing and eliciting specific desired actions. Unless one wants to claim that Batteau's simple methods accessed an otherwise untapped part of the already-complex marine mammal's communicative life, it then seems reasonable, if not necessary, to conceive of native dolphin noises as symbolically meaningful as well. In fact, I am almost coming to the point where I will call them speech acts.

The last type of animal communication that I will mention in this section is, unsurprisingly, primate communication. Considerable attention has been given to ape-human communication, although such discussions are often couched in evolutionary terms that serve to deepen the perceived communicative divide between humans and animals by highlighting an alleged symbolic turning point humans passed. The native calls of Vervet Monkeys, however, demonstrate the symbolic qualities of primate speech acts. As with starlings, these animals have been recorded as having what one might call an identifiable *language* of warnings. According to William Roberts, distinct alarm calls are raised by the monkeys dependent upon the type of predator that is detected nearby: "Short, tonal calls [are] emitted in the presence of a leopard, low-pitched staccato grunts [are] made when an eagle [is] seen, and high-pitched chutters [are] emitted if a snake [is] detected" (361–362). The responses to these symbolic calls—calls that colloquially refer to a concept that must be understood within a particular context by a particular group who knows the language—are very specific. For the leopard call, the Vervet response is to climb high into the trees. The monkeys scan the skies and run into the bush when the eagle call is given. In the instance of the snake call, the monkeys coalesce on the located snake in a mob-like fashion in order to subdue the predator.

In the studies of these monkeys, the calls were recorded and played back to test whether the calls had symbolically motivated effects or were merely part of a reactionary milieu caused by predator presence. One might suggest that the responses were merely monkey-see/monkey-do (literally), where one Vervet Monkey sees a leopard, screams and then runs to the tree tops to avoid predation, followed by the rest of the troop who saw the first monkey running and merely mimicked it. In the playback tests, however, the warning sound was emitted by a speaker system near a group of monkeys. The monkeys all responded to each distinct call in the predicted way, even in the absence of either a calling monkey to mimic or an actual predator (Roberts 362). Thus, it was the referential content of the audible message that elicited the reaction from members of the community.

In 2009, a study by Schel, Tranquilli and Zuberbühler expanded the Vervet exemplar, demonstrating that a number of species of monkeys have been scientifically recorded as having distinct call systems laden with discrete meaning and that they can recognize the particular calls of other species (146–147). Playback experiments were conducted with King Colobus and Guereza Colobus monkeys in which typical vocalizations of leopards and crowned eagles were played within the vicinity of different monkey groups. Vocalization responses from the monkeys were recorded and the researchers observed that "both species [of monkey] produce two basic alarm call types, snorts and acoustically variable roaring phrases, when confronted with leopards or crowned eagles" (136). The researchers found that there were "strong regularities in call sequencing. Leopards typically elicited sequences consisting of a snort followed by a few phrases, while eagles typically elicited sequences with no snorts and many phrases" (136). The vocalizations were meaningful to the community only because the appropriate grunts and snorts were uttered, not because monkeys simply instinctually respond to any loud noise. This is a rather clear actualization of the animals' linguistic capacities.

Were I explaining a dialect of some foreign human tongue, I might rely on the language of semiotic forefather Saussure to describe the example above, instead of relying only on the casual language of *grunts* and *snorts*. Indeed, these grunts are more than just grunts. The groups being studied were uttering particular vocally constructed *phonemes*, or compilations of sounds produced by the vocal cords. These phonemes were combined into *morphemes*, or particular groupings/constructions of phonemes, forming what are typically simply called *words* (Saussure 65–68). These words were then combined into larger semiotic units that pointed referentially to an external object. Using the language of Austin, these symbolic references vocalized by the Vervet and Colobus Monkeys were locutionary acts. They also consisted of an illocutionary warning within the calls creating the promise of imminent danger. And finally, these calls, these symbolic speech acts, were designed to elicit a particular perlocutionary response from those within earshot.

To make clear what I am saying, the language of communication theory fits almost seamlessly into these examples of animal communication when one allows it to. This not only supports the notion that animals engage in what is known as "symbolic communication," but it also demonstrates that the communication studies discipline is more prepared for analysis of animal/animal-human communication than it might realize.

USAGE: SEEING THROUGH THE TEARS

Even if the examples above seem to fit within a communication model of symbolicity, it may not be enough to fully destabilize the pseudo-Burkean separation that pervades the field. To "prove" that animals are symbol using, one must demonstrate their capacity for use—for agency/cognition/

consciousness.[6] I must admit, though, that my intention is not actually to prove anything here. Such proof is, within most perspectives, impossible to generate. The goal in this section is instead to demonstrate a reasonable likelihood of animal agency and thus the appropriateness of considering animal communication symbolic, all in an effort to open up the now-closed study of animal/animal-human communication for those within the traditional humanities communication field.

The difficulty of the search for proof that animals are conscious highlights the point that in the scope of this essay, I could not even irrefutably prove that other humans have active cognitive agency. This is because, as Marian Dawkins explains, "we are always on the inside [of our own skins] and can know about the other 'I's' only from the outside of their skins" (1). So instead, we base our belief of others' agency on observations of them in our world and we then act accordingly. A friend appears to be sad, so we comfort him or her. A student appears to be struggling to understand a concept, so we help him or her comprehend. We take for granted that other humans have their own subjective agency because they act in ways that correspond to our experiences of our own subjectivities. For Dawkins, this assessment of external agencies happens on two levels. First, individuals use their set of experiences to predict that other individuals' experiences are similar to their own. Second, they take the particular circumstances that surround the others' experiences into consideration in order to account for the differences that may be between them. There is a gap, though, between humans' experiences and animals' experiences. As Dawkins recognizes, "our knowledge of experiences inside other bodies is strictly limited. The study of consciousness in other animals would therefore seem doomed to fail" (10). Bridging this gap thus carries with it the real risk of anthropomorphism—trying to make animals fit our own human shape. Dawkins' second level of understanding, where one takes the particular circumstances into account, should theoretically ameliorate fears of problematic anthropomorphism. For sociologist Leslie Irvine, the difference is really between anthropomorphizing and *over*-anthropomorphizing. According to Leslie Irvine, a healthy *critical anthropomorphism* acknowledges that anytime we try to understand any other, be it an animal or another human, we anthropomorphize them inasmuch as we can only see them through our own subjective human experience. This then provides a basis for understanding animals via a "middle ground." As Gordon Burghardt writes,

> Critical anthropomorphism aims to do for the understanding of animal life what *Verstehn* [sic] (Weber 1949) tries to capture in human life, which is to understand the meanings that people give to their actions. *Verstehn* [sic] involves placing oneself in the position of another person to see what purpose his or her actions might have, or, more accurately, to see what that person believes his or her actions will accomplish. Critical anthropomorphism tries to do the same for the experience of

nonhuman animals. Bekoff (2002: 48) refers to it as "humanizing animals with care," for it respects the "natural history, perceptual and learning capabilities, physiology, nervous system, and previous individual history" of animals. (Burghardt 5)

Animal tears exemplify the need to "take care" when looking at animal cognition. Tears are produced under a number of different situations. One might want to say that when a baby seal is seen with a tear in its eye, it must be feeling sad. That, however, would be a care*less* reading of the situation. Even in humans, tears are not a necessary indication of emotion or cognition as a speck of dust can stimulate the lacrimal (tear) glands. It is this sort of quick reading that could warrant the critiques of anthropomorphism and should thus be resisted as studies of animal symbolicity are taken up by communication scholars. Yet, with care, one might still find that animals do at times act with nuanced consciousness. Even Charles Darwin, who held that emotional tears were almost entirely a special expression of humans, hinted at the possibility of animals exhibiting emotions through tears, writing but not refuting that "in the Zoological Gardens the keeper of the Indian elephants positively asserts that he has several times seen tears rolling down the face of the old female, when distressed by the removal of the young one" (168). The task for the contemporary researcher is to proceed with caution, keeping open the possibility of animal cognition without jumping headlong into false-positives. For Irvine, what is needed is an expanded model of animal selfhood. We can understand other humans through their symbolic and emotional activities, and the presence of cognitive capacities in animals can be understood in similar ways. Irvine's expanded model then sees animals as "subjective others" with "agency, coherence, affectivity, and history" (3). Anecdotal evidence, like that given to Darwin by the zookeeper, is thus given new theoretical force when one allows it to be assessed through a critical lens of subjective awareness.

A key issue in assessing these anecdotes is of course the notion of *instinct*. Instinctual actions are certainly present within the animal world. These behaviors are those that are unlearned, naturally occurring tendencies that lead animals toward acts of self-preservation. Analysis must then be opened up to see if there are other un-anomalous actions taken by animals that go against or exist independently of natural, unlearned impulses toward survival. Instances where animals act in meaningful ways opposed to self-preservation may then point to a higher level of conscious cognitive ability than is typically acknowledged.

Such is the case with a pair of Peregrine Falcons, Arthur and Jenny, who were successfully parenting five fledglings. Biologist Marcy Cottrell Houle was observing the pair when one day Jenny went missing. According to Houle, after Jenny did not return, Arthur's behavior changed dramatically. For the first two days she was missing, Arthur continued to meagerly attend the fledglings while continually calling for his mate. On the third day, with

Jenny still missing, Arthur uttered a sound he had not made before. Houle described it as "a cry like the screeching moan of a wounded animal, the cry of a creature in suffering," adding that, "the sadness in the outcry was unmistakable; having heard it, I will never doubt that an animal can suffer emotions that we humans think belong to our species alone" (77). Arthur then spent the fourth day motionless on a nearby rock. The fifth day was full of intense hunting, bringing food to the fledglings, as Arthur's actions became more moderated. Houle noted that while two fledglings had survived and fledged successfully, three had died in the ordeal. No simple self-preservation or natural-reaction explanation wholly accounts for this. It seems reasonable to consider that perhaps Arthur's actions came from a place of emotive agency instead of mere instinct.

Similarly, the well-known herd-survival practices of elephants may not strike one as a demonstration of consciousness, but the extent to which elephants have been observed caring for herd members speaks volumes about the likelihood of higher levels of elephant awareness. In addition to using their trunks to carry dying herd members, elephants have been observed trying to feed other elephants too sick or injured to feed themselves. Like Arthur the falcon, elephants are also known to act in ways contrary to personal or even herd survival. Joyce Poole and Petter Granli have plainly labeled one such elephant activity as *body guarding*, the risky behavior of an elephant standing over the body of a dead elephant or human to protect the dead body from predators or other apparent threats (123–124). As Jeffrey Masson and Susan McCarthy explain, distilling the extensive work of ethologist Cynthia Moss, these types of actions suggest that elephants are deeply affective, perhaps even grieving over loss, particularly evidenced by elephants' observed fascination with bones of their own species. When they find elephant bones, they often inspect them and even carry them for miles, but this behavior does not occur with respect to the bones of other species (95–96).

Many instances of perceived animal pleasure have been recorded as well. Irvine describes her cats' love for their blanket in a way that mirrors many stories animal enthusiasts share. For Irvine's cats, the joy a blanket brings them at home, though, does not quite match the unsettledness the veterinarian's office brings, pointing to the cats' distinctly contextual emotional awareness (14–15). Animals in the wild also seem to have their own moments of joy amidst their struggle for survival. Before recorded distress calls and tame falcons were used to scare them off, crows inhabited the areas around the gold leaf domes of the Kremlin (they were purposefully scared off because of the damage they were causing). They had found it enjoyable to slide down the onion-shaped domes for no apparent purpose, their claws doing the damage on the way down. Bears have also been seen sliding down snow banks like otters. Other bears have been seen floating in mountain lakes, sticking their noses under water to blow bubbles, and then using their claws to pop the bubbles (Masson and McCarthy 122–129). An explanation of these activities as pure instinct should strike the critically open investigator as potentially shallow and prejudiced. One could of course argue that play is a part of learning how to defend oneself. The rule of parsimony, however, seems to side with an enjoyment theory and not a convoluted practice-bursting-bubbles-is-necessary-for-survival theory. Even within survival strategies, though, cognitive ethologist Donald Griffin contends that one can find evidence suggestive of animal consciousness given the creativity of altered routine behavior:

> In a city park in Japan, a hungry green-backed heron picks up a twig, breaks it into small pieces, and carries one of these to the edge of a pond, where she drops it into the water. At first it drifts away, but she picks it up and brings it back. She watches the floating twig intently until small minnows swim up to it, and she then seizes one by a rapid thrusting grab with her long, sharp bill. Another green-backed heron from the same colony carries bits of material to a branch extending out over the pond and tosses the bait into the water below. When minnows approach this bait, he flies down and seizes one on the wing.
>
> Must we reject, or repress, any suggestion that the chimpanzees or the herons think consciously about the tasty food they manage to obtain by these coordinated actions? Many animals adapt their behavior to the challenges they face either under natural conditions or in laboratory experiments. This has persuaded many scientists that some sort of cognition must be required to orchestrate such versatile behavior. (Griffin 1)

These innovative actions, which demonstrate only a portion of animal mindfulness, are products of thought that is cogently purposive and contextually inventive—that is, awareness sufficient for the use of symbols.

Although a full reworking of any particular consciousness theory, such as Irvine's expanded model of animal selfhood, is not fleshed out in these examples, these few brief anecdotes of animal behavior do seem to point to some dimension of their consciousness beyond mere instinctual mechanization. The simplistic, neurological, evolutionary descriptions of animal-automatons are insufficient and inaccurate. It is not my intention to advance a perfected theory of animal consciousness. The question is also not whether animals and humans operate on the same logic(s)—although at times they may. Instead, my goal is simply to make plausible the belief that animals *use* the symbols they utter—that they act with a certain degree of purposive agency when they communicate.

CONCLUSION: TOWARD OPENNESS AND TRANSDISCIPLINARITY

To return to the larger issue, the point here is that Burke's "Definition of Man," oft recited as simply "Man is the symbol-using animal," is fundamentally flawed. It operates under its own perfective-rottenness and serves to create a clear and distinguishable communicative divide between humans and animals that simply does not exist. The evidence indicates that this exclusionary model is wrong, as animals do consciously symbolically communicate.

I do not mean to say that there are no differences. Such a position seems as inherently naïve as the artificial binary that has been established as absolute between animal and human communications. In response to the contentions put forth in this essay, I often hear the counterargument that animals do not engage in metacommunication. I most frequently hear this in the form of "*Well, you don't see animals writing papers about their communication, now do you?!*" I must concede that there is truth in that reductive statement, in that humans spend much more time articulating the arts and sciences of our own communicative habits than is evidenced even slightly anywhere in animal behavior. Yet, that argument has very little impact on the position articulated here. The conclusion is not that there are no differences between animal and human communication, but, as the saying goes, the differences are largely in degree, not type. Whereas animals do not engage in communication to the degree of complexity that humans often do, they still go through processes of learning and refining their symbolic communication. Dolphins, for instance, have a complex symbolic system which some believe must be taught to the young dolphins, usually during the first twenty months that the dolphin spends with its mother before being weaned (Evans 132). Capuchin Monkeys, another species with complex communicative abilities, often give false or misguided warning calls when they are young, sounding calls that mean nothing or that signal the wrong response to a predator nearby. Sometimes these erroneous calls serve as a form of deception in order to keep the troop away from a prized food treat. As they get older, the monkeys must learn the correct cries from the rest of the troop—sometimes learning through censure, which "implies that they're beginning to think about what each other may be thinking about" ("Clever Monkeys").

The broader consideration of metacommunication in animal communities is fascinating, but truthfully it has little to do with the core critique offered in this chapter. Even if it were concluded that only humans engage in metacommunication, it would not change the fact that the disciplinary reliance on a symbolicity schism between human and animal communication is based on a fundamentally flawed principle, and that it has foreclosed fascinating avenues of research for too long. Surely not all human communication is meta-oriented. Most of it is not. Yet it is still the use of symbols, and it is still meaningful and valuable subject matter for the communication discipline. In fact, it is easily the most common subject matter for the communication discipline. That animals might not engage in metacommunication thus seems to be no reason to exclude animal and animal-human communication from the field's purview. To do so would not only be inconsistent, but would also invoke a level of hubris that fails to acknowledge our own incapacity to know or control language. As Condit said, one should "revise Burke's label to the 'symbolizing animal' in order to reflect feminist and post-structuralist emphasis that humans are as used by their symbols as they are users of those symbols" (8). My amendment to her insightful statement would be that the definition, if kept at all, should be revised to say that humans are *a* symbolizing animal.

The irony in all of this is that elsewhere in his thinking Burke himself was not nearly as dogmatic as his definition appears to be. On the question of animal action, though, Burke was not as Burkean as one might hope, a limitation contributing to the strident separation between the disciplines that study human and animal communication. It is very rare to find an article addressing animal or animal-human communication in any major communication publication. Those that do get published rarely have much heuristic aftereffect, which I contend is in large part the result of disciplinary biases modernly rooted in and emblematized by Burke's definition. Rarely do you see the humanities and ethology fields interacting with one another, holding joint conferences, or even being housed anywhere near each other on college campuses. This, of course, is not all the humanities' fault. Those in scientific fields routinely look down upon the "soft" sciences like communication studies. Their publications even entrench the division, using language such as "biological communication" (e.g., Scott-Phillips), and mechanical terms like "signal reliability" instead of "deceit" or "lying" (e.g., Searcy and Nowicki). This sterilization of the language, which Griffin refers to as "semantic piracy" (5–6), serves to remove any trace of mystique from the natural world. Communication is typically seen as reducible and knowable, instead of open or uncertain. There have been some noteworthy exceptions to this tendency in both fields. Classicist George Kennedy's "A Hoot in the Dark" (1992) provides an exciting and unexpected look at the rhetorical aspects of communication in nature. Sadly, he backs off from his claims at the end of the article, making animal communication merely an analogy that can point to truths about human rhetoric. Although his explanation of animal rhetoric is impressive and visionary, the anthropocentric ending is unfortunately rather routine. Across the aisle, Pederson and Fields offer a useful demonstration of integrative communication work, using discourse analysis to explain the data they collected on conversations between Bonobo Apes and humans, concluding that the Bonobos' repetition served a linguistic function, co-constructing the conversational turns. Their article, however, is part of a rare breed.

What is needed then is a new spirit of openness and transdisciplinarity. As Debra Hawhee has compellingly laid out, this is in fact the approach that Burke's work should inspire. Burke's definition may routinely form part of a narrow view of communication, but Burke's larger thinking about communication is vast and nuanced concerning language, the body, movement and cognitive and noncognitive beginnings. For Hawhee, Burke's work demonstrates a deep dissatisfaction with a limited understanding of communication, instead embracing what was truly a more vast and inclusive approach to inquiry. She writes, "From an early point in his career and however unwittingly, Kenneth Burke became something of a spokesperson for transdisciplinarity" (3). This transdisciplinarity is not merely information or topic sharing, but rather requires a new way of seeing:

> What distinguishes transdisciplinarity from interdisciplinarity is its effort to suspend—however temporarily—one's own disciplinary terms and values in favor of a broad, open, multilevel inquiry. . . . The difference is a matter of sharing methodologies (something interdisciplinarity does quite well) versus broadening perspective, one of the main goals of transdisciplinarity. (Hawhee 3)

Transdisciplinarity is an approach that requires "a deliberate forgetting of what one already knows" in favor of new possibilities between "their discipline" and "ours" (Hawhee 4). The goal is not to redefine *symbol*, to rest on differences in metacommunication or even to be satisfied with the use of classic communication theorists like Austin or Saussure. The project here is to open up and justify that new space—a new perspective on our inquiry—by allowing for discourse on human, animal and human-animal communication across disciplinary boundaries, however these boundaries have come about.

The communication studies field is rich with provocative, meaningful and impactful studies. The discipline's artificial casting of animal communication as wholly foreign and unsymbolic, however, has perpetuated a division that robs the discipline of further riches. This separation has been prolonged by the recitations of Burke's definition and evidenced in the content and style of the scholarship at all levels of the field. To unsettle such normative habits may require the occasional un-defining, but more importantly it requires a willingness from the members of the field to see the rich possibilities in the paths of serious scholarship that have yet to be explored. They are paths about which the field should be excited.

NOTES

1. This project is one that I began as an undergraduate student at Liberty University after reading Kenneth Burke's "Definition of Man" in a rhetorical theory course with debate coach Brett O'Donnell. It has been a project that I have explored at almost every step of my academic career since then. Though the field of communication studies has many normative boundaries, I owe many thanks to many individuals (supporters and dissenters alike) for helping this project develop. I especially owe thanks to Faith Mullen and Lynnda Beavers of Liberty University who, through the work of my undergraduate honors thesis, encouraged me to see the field through a different and open lens. Later, during my doctoral studies, perpetual openness and encouragement from the sophistic thinker Victor Vitanza, along with a course on Burke from expert David Blakesley, propelled this work to new depths for which I am grateful.

2. Burke's "Definition of Man" appeared first in *The Rhetoric of Religion: Studies in Logology* (1961) and then in an article in *The Hudson Review* (1963–1964); its most-cited appearance is in *Language as Symbolic Action* (1966).

3. I certainly acknowledge the gendered implications of the term "Man," yet such an exploration is the property of other inquiries and lies outside the scope of this essay. Having said that, it does seem likely that such gendered thinking may also lie at the root of the human-animal schism inherent in most communication scholarship. I will endeavor to avoid reinscribing the term in the essay, but will nonetheless use it when Burke did in order to preserve the original text in its own form.

4. Whereas Burke largely spoke in ways consistent with a notion of free will, the question of *conscious choice* pervades his explanations in *Permanence and Change*. Poetic force and idiosyncratic synthesizing of interpretations may indeed give humans choice (65–66, 229, 257 footnote 2). Yet orientation, ideology and terministic language have ways of seeming to compel action in a deterministic sense and thus humans might not have free agency (5–18, 233–234). It may be the case, Burke contends in *The Philosophy of Literary Form*, that the critic has to adopt both a deterministic and free-will perspective, even if doing so results in inconsistency (115–116). Perhaps most pragmatically, Burke explains, even if one were to conclude that choices are determined, humans still act as if they have choice (*Permanence and Change* 218–219), allowing critics to ask further questions about motives.

5. See *Counter-Statement* (53); *Grammar of Motives* (135–137); *Philosophy of Literary Form* (8–9).

6. I am using terms like *conscious, sentient, aware* and *cognitive* in overlapping and interchangeable ways. That is not because I see these terms as equivalent, but because they all hint at the level of agency that I am arguing animals demonstrate. Different theorists describe this in different ways (such as Irvine's use of the term *subjective others*), and their variations are useful, but the semantic differences are beyond the scope of this essay. Moreover, Donald Griffin suggests avoiding concrete definitions of the term given our limited understanding of the phenomenon, regardless of which species is being discussed—human or not (4–5).

REFERENCES

Austin, J. L. *How to Do Things with Words*. 1975. 2nd ed. Oxford: Oxford UP, 2003. Print.

Baldwin, John R., Stephen D. Perry, and Mary Anne Moffitt. *Communication Theories for Everyday Life*. New York: Pearson, 2004. Print.

Bekoff, Marc. *Minding Animals: Awareness, Emotion, and Heart*. Oxford: Oxford UP, 2002. Print.

Burghardt, Gordon M. "The Evolutionary Origins of Play Revisited: Lessons from Turtles." *Animal Play: Evolutionary, Comparative, and Ecological Perspectives*. Ed. Marc Bekoff and John Alexander Byers. Cambridge: Cambridge UP, 1998. 1–26. Print.

Burke, Kenneth. *Counter-Statement*. 1931. Berkeley: U of California P, 1968. Print.

———. "Definition of Man." *Hudson Review* 16.4 (1963–1964): 491–514. Print.

———. *A Grammar of Motives*. 1945. Berkeley: U of California P, 1969. Print.

———. *Language as Symbolic Action*. Berkeley: U of California P, 1966. Print.

————. *Permanence and Change*. 3rd ed. Berkeley: U of California P, 1984. Print.
————. *The Philosophy of Literary Form*. 3rd ed. Berkeley: U of California P, 1973. Print.
————. *The Rhetoric of Religion: Studies in Logology*. 1961. Berkeley: U of California P, 1970. Print.
"Clever Monkeys." *Natural World*. BBC One. London. 25 Nov. 2008. Television.
Condit, Celeste M. "How Should We Study the Symbolizing Animal?" *The Carroll C. Arnold Distinguished Lecture, National Communication Association*. Nov. 2004, Chicago, IL. Boston: Pearson Academic, 2006. Print.
Crist, Eileen. *Images of Animals: Anthropomorphism and Animal Mind*. Philadelphia: Temple UP, 2002. Print.
Darwin, Charles. *The Expression of the Emotions in Man and Animals*. 1872. Oxford: Oxford UP, 1998. Print.
Davis, Flora. *Eloquent Animals*. New York: Coward, McCann & Geoghegan, 1978. Print.
Dawkins, Marian Stamp. *Through Our Eyes Only?* New York: Oxford UP, 1998. Print.
Evans, William. *Communication in the Animal World*. New York: Crowell, 1968. Print.
Fisher, John Andrew. "Disambiguating Anthropomorphism: An Interdisciplinary Review." *Perspectives in Ethology*, vol. 9: *Human Understanding and Animal Awareness*. Ed. Patrick Bateson and Peter Klopfer. New York: Plenum, 1991. 49–85. Print.
Gammon, David E., Melinda C. Hendrick, and Myron C. Baker. "Vocal Communication in a Songbird with a Novel Song Repertoire." *Behaviour* 145 (2008): 1003–1026. Print.
Griffin, Donald R. *Animal Minds: Beyond Cognition to Consciousness*. Chicago: U of Chicago P, 2001. Print.
Griffin, Em. *A First Look at Communication Theory*. 6th ed. New York: McGraw-Hill, 2006. Print.
Hardt, Hanno. *Critical Communication Studies: Communication, History, and Theory in America*. New York: Routledge, 1992. Print.
Hawhee, Debra. *Moving Bodies: Kenneth Burke at the Edges of Language*. Columbia: U of South Carolina P, 2009. Print.
Henderson, Greig. "Postmodern Burke." *University of Toronto Quarterly* 66.3 (1997): 562–575. *Academic Search Premier*. Web. 5 May 2010.
Houle, Marcy Cottrell. *Wings for My Flight: The Peregrine Falcons of Chimney Rock*. Reading, MA: Addison-Wesley, 1991. Print.
Irvine, Leslie. "A Model of Animal Selfhood: Expanding Interactionist Possibilities." *Symbolic Interaction* 27.1 (2004): 3–21. *JSTOR*. Web. 5 May 2010.
Kennedy, George A. "A Hoot in the Dark: The Evolution of General Rhetoric." *Philosophy and Rhetoric* 25.1 (1992): 1–21. *JSTOR*. 5 May 2010.
Littlejohn, Stephen W., and Karen A. Foss. *Theories of Human Communication*. 7th ed. Belmont, CA: Wadsworth, 2002. Print.
Masson, Jeffrey Moussaieff, and Susan McCarthy. *When Elephants Weep*. New York: Dell, 1995. Print.
Mitchell, Robert, Nick Thompson, and Lyn Miles, eds. *Anthropomorphism, Anecdotes, and Animals*. Albany: SUNY P, 1997. Print.
Moss, Cynthia. *Elephant Memories: Thirteen Years in the Life of an Elephant Family*. Chicago: U of Chicago P, 2000. Print.
Murray, Jeffrey. "An Other Ethics for Kenneth Burke" *Communication Studies* 49 (1998): 29–45. Print.
Owings, Donald H. *Animal Vocal Communication: A New Approach*. New York: Cambridge UP, 1998. Print.

Pederson, Janni, and William M. Fields. "Aspects of Repetition in Bonobo-Human Conversation: Creating Cohesion in a Conversation between Species." *Integrative Psychological and Behavioral Science* 43 (2009): 22–41. *Academic Search Premier.* Web. 5 May 2010.

Poole, Joyce H., and Petter Granli. "Signals, Gestures, and Behavior of African Elephants." *The Amboseli Elephants: A Long-Term Perspective on a Long-Lived Mammal.* Ed. Cynthia Moss, Harvey Croze, and Phyllis C. Lee. Chicago: U of Chicago P, 2011. 109–124. Print.

Proppe, Darren S., and Gary Ritchison. "Use and Possible Functions of the Primary and Sustained Songs of Male Grasshopper Sparrows." *The American Midland Naturalist* 160.1 (2008): 1–6. *Academic Search Premier.* 5 May 2010.

Roberts, William A. *Principles of Animal Cognition.* Boston: McGraw Hill, 1998. Print.

Rossong, Melanie, and John Terhune. "Source Levels and Communication-Range Models for Harp Seals (*Pagophilus groenlandicus*) Underwater Calls in the Gulf of St. Lawrence, Canada." *Canadian Journal of Zoology* 87.7 (2009): 609–617. *Academic Search Premier.* 5 May 2010.

Sanders, Clinton R. *Understanding Dogs: Living and Working with Canine Companions.* Philadelphia: Temple UP, 1999. Print.

Saussure, Ferdinand de. *Course in General Linguistics.* 1916. Trans. Wade Baskin. New York: McGraw-Hill, 1966. Print.

Schel, Anne Marijke, Sandra Tranquilli, and Klaus Zuberbühler. "The Alarm Call Systems of Two Species of Black-and-White Colobus Monkeys (*Colobus polykomos* and *Colobus guereza*)." *Journal of Comparative Psychology* 123.2 (2009): 136–150. *PsycARTICLES.* Web. 5 May 2010.

Scott-Phillips, Thom. "Defining Biological Communication." *Journal of Evolutionary Biology* 21 (2008): 387–395. Print.

Searcy, William A., and Stephen Nowicki. "Bird Song and the Problem of Honest Communication." *American Scientist* 96.2 (2008): 114–121. *Academic Search Premier.* Web. 5 May 2010.

Weber, Max. *The Methodology of the Social Sciences.* Glenco, IL: Free Press, 1949. Print.

14 Difference without Hierarchy

Narrative Paradigms and Critical Animal Studies—A Meditation on Communication

Susannah Bunny LeBaron

IN THE BEGINNING

Gardening is, generally, considered the sport of gentle men and women in floppy hats. Soothing, moderate and meditative, it is recommended by psychologists and physicians alike as a means to cultivate health and relaxation. To celebrate this peaceful and uplifting pastime, many decorate their horticultural endeavors with statues of St. Francis, the eco-saint, holding out his hands for birds to alight and peck at seed. You will also frequently find the words of Dorothy Frances Gurney emblazoned on a plaque and hung upon a fence or trellis: "The kiss of the sun for pardon, / The song of the birds for mirth, / One is nearer God's heart in a garden / Than anywhere else on earth."

Yet also in the garden one finds the bees—buzzing sensuously over every penile filament of the flowers—and the birds—the cascade of their melodious notes accompanied by the tearing of worms from the earth, the raiding of one another's nests and the tumbling about of avian sexcapades. Further, as a gardener, one is constantly making decisions about who lives and who dies—which plant, which insect—and these decisions are often ruthlessly executed. Our Western narrative of gardening is rooted firmly in the Abrahamic story of Eden, and that is a story of perfection, a moment of primordial oneness with our creator. Adhering to these narratives of the garden as perfection and the gardener as divine provides a great deal of psychological remove from the realities of biological life that surround us in a garden. Furthermore, those narratives allows us to continue our daily lives—whether or not we actually garden—within naturalized patterns of behavior that are predicated on the belief that we, as human beings, are something different from the rest of the earth's biota. We are set apart from—above—our ecologies, but, according to our narratives, that's because we're special.

In the following meditation, I consider examples of this narrative of separateness and superiority, as well as challenges to it. I begin with my own experiences as a gardener, and I use Fisher's theory of *Homo narrans* to examine it, as well as: a speech by famous primatologist Dr. Robert Sapolsky, recent work on plant-insect interaction, medical research and Derrida's poignant questions about differences between humans and animals. Meditation emerges from the combination of receptivity and focus. Often, a chime or bell is struck to mark the beginning of that psychic space. Consider the following, then, a chime, a sounding out of the space to come, two thought-bells; they are brief, yet complete and, as Claude Lévi-Strauss said about animals, they are "good to think with."

GOOD TO THINK WITH

One: A boundary is more like a membrane than a wall. (Conquergood 145)
Two: A few years ago my friend Rod and I were riding our bicycles around Boulder, Colorado, when we witnessed a very interesting encounter among five magpies. Magpies are corvids, a very intelligent family of birds. One magpie had obviously been hit by a car and was lying dead on the side of the road. The four other magpies were standing around him. One approached the corpse, gently pecked at it—just as an elephant noses the carcass of another elephant—and stepped back. Another magpie did the same thing. Next, one of the magpies flew off, brought back some grass, and laid it by the corpse. Another magpie did the same. Then, all four magpies stood vigil for a few seconds and one by one flew off. (Bekoff 1)

THINGS DO NOT GO AS PLANNED

I grew up with gardeners, but my own work with domesticated flora has been unsteady. A few pots here, a few bulbs there. Nothing serious. Then my partner and I bought a house and ripped out the front lawn, and I began planting. It was in the process of caring for the flowers and, especially, the vegetables and melons I was growing, that I came to understand just how fraught this "gentle" pastime really is.

It was early August, and I had not been in my garden in days. Not really. I had been out to water and to pick some tomatoes, but I had not given it the serious attention of weeding or trimming—or even just gazing at it as I planned for next year. So, imagine my horror when I finally began tending to the cucurbits patch and discovered my zucchini plants *covered* with the little brown and grey bodies of *Anasa tristis*. There were nymphs of all sizes— just hatched to almost grown. There were adults with their brown-and-black backs that make them harder to see. There were clusters of carnelian-colored egg deposits. And I was aghast. *Aghast.* I threw on my gloves and begin killing them, squishing their little bodies. But there were too many. I went inside to mix up a strong batch of neem-and-dish-soap insecticide (pollinator friendly) and then I sprayed and sprayed and sprayed. As they were knocked from their perches, I ground them into the dirt to make sure that they died.

Of course, the neem doesn't actually kill them. That's the reason I use it. Any insecticide that would have quelled my bloodlust by killing instantly is bad for the bees and the moths that do gardenly good deeds. Yes, even the "eco" sprays that use cedar and peppermint oil.[1] The neem acts only on the bugs I want to kill, but it just sterilizes them. Or makes them forget to eat or mate. But it only works on the bugs in the nymph stage, so you'd better believe I just *soaked* any little grey bodied *Anasas* that I couldn't actually step on.

Their common name, as fate would have it, is "squash bugs," and they're a problem almost everywhere that people grow cucurbits (squashes, melons, gourds, etc.). *Mother Earth News* proposes that you lay out boards and flip them over in the morning to squish the bugs who have taken shelter there. "We're urging readers to test this method of using a 'squash bug squisher' made of two boards hinged together, which makes the 'squishing' part easier" (Long 105). But I hadn't had time to hinge boards. So my killing was mostly individual; it felt brutal and visceral. It frightened me. Because the fact of the matter is that, if we didn't have conflicting agendas regarding zucchini plants, I'd have found them cute. The newborns are all spindly legged and they sort of hang out in a group as if unsure of the big world. And the older nymphs are the softest shade of dove gray. The adults have a tiny, neat, convex curve between their heads and their shoulders, and their coloring is so clever—it's like bark or dirt—and I admire clever in all its forms. Even the egg deposits are beautiful; they're small clusters of perfectly round, perfectly spaced eggs, usually deposited under a leaf near a branching in the veins, so they're protected. Until I came along.

That day, all the stages of the *Anasa tristis* were on display. They were gathered in multiple generations as if at a giant family reunion, or a jamboree. So even as I knocked them to the ground or stepped on them or squished them with my gloved hands or drowned them in the environmentally friendly slow death of neem oil, I was quite aware, painfully aware, that I was acting very much like a gardener—but I did not feel divine. I felt a little like a Nazi. And this is why.

HOMO NARRANS AFTER ALL

We are, taxonomically speaking, wise. We walk erect and have large brains, and thus we have named ourselves *Homo sapiens*. "Many different root metaphors have been put forth to represent the essential nature of human beings," writes Walter Fisher, and he goes on to list some, including: *Homo faber,* 'man the maker,' who controls the environment through his tools; *Homo economicus*, 'man the rational and economically self-interested,' who gets what he can for as little as he can; *Homo sociologicus*, 'man the environmentally determined' who, like a *tabula rasa*, is writ on by the society in which he is raised (62). To this list, actually *over* this list, Fisher wants to add *Homo narrans*:

> When narration is taken as the master metaphor, it subsumes the others. The other metaphors become conceptions that inform various ways of *recounting* or *accounting for* human choice and action. Recounting takes such forms as history, biography, or autobiography. Accounting for takes such forms as theoretical explanation or argument. . . . Regardless of the form they are given, recounting and accounting for constitute stories we tell ourselves and each other to establish a meaningful life-world.
>
> . . . The idea of human beings as storytellers posits the generic form of all symbol composition. It holds that symbols are created and communicated ultimately as stories meant to give order to human experience and to induce others to dwell in them in order to establish ways of living in common, in intellectual and spiritual communities in which there is confirmation for the story that constitutes one's life. . . . As Heidegger observed, "We are a conversation . . . conversation and its unity support our existence." (62–63)

And that conversation, that collectively woven story, is the lens through which we view the world, consciously or unconsciously. Consider the parable of the fig tree . . . and the fig wasp.

For eighty million years the fig and the fig wasp have been reproductive partners. The fig tree provides a safe place for the fig wasp to lay its eggs, and the fig wasp pollinates the fig tree. Unless it doesn't. Then "the trees get even by dropping those figs to the ground, killing the baby wasps inside . . . " (Ramanujan). It's like botanical Shakespeare.

Charlotte Jandér, who led a recent Cornell–Smithsonian Institute study on the mutualism of figs and their wasps, frames it as follows:

> Sanctions seem to be a necessary force in keeping this and other mutually beneficial relationships on track when being part of a mutualism is costly. . . . In our study, we saw less cheating when sanctions were stronger. Similar results have been found among human societies and social insects. It is very appealing to think that the same general principles could help maintain cooperation both within and among species. (qtd. in Ramanujan)

Appealing indeed. It posits evolutionary proof for foreign policy and divorce law that take their cue from Medea. The study claims that:

> Combined with previous studies, our findings suggest that (i) mutualisms can show coevolutionary dynamics analogous to those of 'arms races' in overtly antagonistic interactions; (ii) sanctions are critical

for long-term mutualism stability when providing benefits to a host is costly, and (iii) there are general principles that help maintain cooperation both within and among species. (Jandér and Herre 1481)

Metaphorical comparisons to "arms races" reinscribe evolutionary theory with a bellicose "survival of the fittest" ethos. We may as well be ancient Greeks making sure there's plenty of *agon* in our theater productions. Jandér asks, "What prevents the wasps from cheating and reaping the benefits of the relationship without paying the costs?" (qtd. in Ramanujan). Justifiably vindictive fig trees, that's what.

Words like "cheating" and "sanctions" are laden with power dynamics and a moralistic sensibility of justice as portrayed in the Old Testament. These words tell us more about the worldview of the researchers than they do about the actual motivations of either tree or wasp, who have been at this for millions of years. Perhaps John Durham Peters is right when he claims that "at best, 'communication' is the name for those practices that compensate for the fact that we can never be each other" (268). We make a story because we do not know, perhaps *cannot* know, what is going on around us, and too often the connections we make between observable facts obfuscate the facts themselves. These are the stories we tell (in this case an accounting) that do not serve us well. As Fisher's theory of narrative rationality posits, the stories we live in are powered in large part by the values that we hold (48); the story—or explanation—is a bridge between people or groups of people, and it is when the values tacitly embedded in a story align with the values of the listener or reader that the story is accepted without question. This value alignment allows the story to slip beneath the radar of examination. So if you already believe in a strong military, that cheaters never prosper and that something needs to be done about people using welfare to 'leech off hardworking folk,' then chances are you will look at this study on fig trees and their fig wasps and think 'That makes sense,' without recognizing the basis for your agreement.

Science is supposed to be immune to such foibles; the modern world regards it as the clear, sparkling lens of rationality and fact. That is its unshakable persuasion in this age as it holds the Western world's metavalues of objectivity and reason in its tautological grip. Science doesn't get involved; it stands apart and observes. We know this because, as the tradition of scientific observation dictates, scientists use numbers and the nongendered pronoun "it," and they speak to that which is trackable and repeatable.

Consider Jane Goodall's experience when she went to Cambridge, under the auspices of her mentor, Louis Leakey. Despite extensive field experience, her lack of formal training left her ignorant of the norms of the scientific-academic community. At Cambridge, she was brought to task for not adhering to those traditional markers of scientific objectivity:

There I was criticized for my lack of scientific method, for naming the chimpanzees rather than assigning them a number, for "giving" them personalities, and for maintaining they had minds and emotions. For these, I was told sternly, were attributes reserved for the human animal. I was even reprimanded for referring to male chimpanzees as "he" and a female "she": Didn't I know that "it" was the correct way to refer to an animal? Well, a *nonhuman* animal [emphasis in the orignial]. (xii)

It was her thesis advisor, Robert Hinde, who "taught me to express my common sense but ethologically revolutionary ideas in a way that would protect me from too much hostile scientific criticism. . . . I could not say, 'Fifi was happy,' since I could not *prove* this: but I could say, 'Fifi behaved in such a way that, had she been human, we would say she was happy'!" (xiii). In the late sixties, as Goodall notes, "more and more biologists went into the field and started long-term studies on all manner of animals species . . . " (xiii). Faced with the observable complexity of animal behavior, these biologists had to change some of their assumptions, values and, of course, the stories they presented about who and what these animals are.

Thomas Kuhn has led the way in understanding what these shifts are about. His book *The Structure of Scientific Revolutions*, first published in 1962 and currently in its third edition, is "one of the most cited academic books of all time" (Bird). Paradigms, as defined by Kuhn, are "universally recognized scientific achievements that for a time provide model problems and solutions to a community of practitioners" (x). I would cast this in terms of Fisher's theory by stating that the governing narrative changes. In particular, it changes to the degree that it includes the "metavalue" of real-world engagement,[2] that is, the inclusion of available facts; and as the story changes, so too do all of the attendant and implied relationships within it. So it is not just that a major scientific discovery changes the way the data is viewed—Kuhn distinguishes this as interpretation (121–123)—but rather the paradigm facilitates a new story through which we interpret the world, and by which we fashion our constitutive conversations and communities (Fisher 63). However one frames it, the shift is profound:

Led by a new paradigm, scientists adopt new instruments and look in new places. Even more important, during revolutions scientists see new and different things when looking with familiar instruments in places they have looked before. It is rather as if the professional community had been suddenly transported to another planet where familiar objects are seen in a different light and are joined by unfamiliar ones as well. In so far as their only recourse to that world is through what they see and do, we may want to say that after a revolution scientists are responding to a different world. (Kuhn 111)

The difficulty with changing the zoological paradigm is that the story we tell of animals bumps right up against the story we tell about human beings. This means that the story changes slowly, one word at a time: he, she, tool, mind, culture, family, emotion, language. This tentative and understandably judicious[3] use of such words when applied to animals slowly massages the boundaries of our current narrative rationale, thereby exposing the contour of the governing system of values.[4]

BOUNDARIES, HARD AND SOFT

The 2009 graduating class of Stanford University elected Dr. Robert Sapolsky to deliver the Class Day Lecture, and thus, on June 13 of that year, Sapolsky took the podium and delivered "The Uniqueness of Humans" to a packed basketball stadium. Sapolsky is an extraordinary public speaker; he is engaging, witty and inspirational—if nothing else in the sheer breadth of his knowledge, which he manages to communicate with coherence and enthusiasm.

Sapolsky is a neuroendocrinologist and a primatologist who studies the complexities of stress among primates, including humans. Whereas his primary research subjects for the past thirty years have been a troupe of wild olive baboons in the Masai-Mara National Reserve in southwestern Kenya, his findings illuminate and are applicable to human society as well. He has won numerous awards for his research, his teaching and his popularization of science through his books *Why Zebras Don't Get Ulcers, The Trouble with Testosterone* and *A Primate's Memoir,* and more. Even watching him on YouTube, one can clearly see why Stanford students would want to hear this man hold forth.

After the introductory comments, Sapolsky introduces his topic by way of a charming misquote:

> What I thought I would do today is talk a bit about how one thus makes sense of humans in this context of "We're a primate—we're just another species," and it will hark back to a famous quote from some evolutionary biologist who said that, "All species are unique, but humans are uniquier" . . . or something like that. . . . what I'll focus on here is what it is that makes us not so unique-ier and ways in which it does.

He begins by pointing out that, if you're looking for what makes humans unique, it's not going to be in our biological building blocks—not our genes or neurons, or neurotransmitters, etc. He states that "sometimes the challenge is recognizing that there's nothing unique about us; we're just a basic, off-the-rack mammal." From there, he moves on to his second category of human uniqueness by degree, wherein we use those basics in "ways that are unprecedented."

Here he goes through a number of different categories, each of them defined by a word or phrase, and each of them traditionally thought to be the exclusive realm of humanity. The categories are: aggression, theory of mind, golden rule, empathy, anticipation and reward, and culture. The exploration of each of these different categories ends in pointing out how—although we share the basics of these behaviors with other species—we take it to a level beyond what they could imagine. The first, of course, is not flattering, as Sapolsky discusses the use of drone bombers in Iraq.

One might have assumed, from the introduction to his speech, that Sapolsky would be engaging in a celebration of the ways in which humans are different from or the same as other forms of life, specifically primates, embracing an eco-ethic wherein we are all an integral part of the web of life, neither higher nor lower than the others, neither sovereign nor sundered, but with and among—part, yes, but part *of*. One might argue that beginning with the pathological height of passive-aggressive behavior that is the remote bombing of a foreign country would have served to hobble any hubris within the audience. But the discussion of the drone bombers functions less as a moral reprimand than to show how *different* we are from other animals. Chimpanzees may have their border patrols and engage in the mass murder of neighboring troupes, but we humans take aggression to a level that has not been seen before. It's the valuing of separateness that shows up here, revealing the narrative that governs Sapolsky's speech, example by example. For, as Fisher notes:

> The concept of narrative rationality asserts that it is not the *individual form* of argument that is ultimately persuasive in discourse. That is important, but *values* are more persuasive, and they may be expressed in a variety of modes, of which argument is only one. Hence narrative rationality focuses on "good reasons"—*elements that provide warrants for accepting or adhering to the advice fostered by any form of communication that can be considered rhetorical* [all emphases in the orignial]. (48)

Sapolsky, his ethos radiating around him like a halo, presents a very convincing list of warrants for accepting the argument that there's nothing like us on earth—and that is what makes us so damn good.

As the lecture continues, we learn that humans are uniquely capable of supposing the thoughts of others in complex relationship involving *more* others, as in Theory of Mind and Secondary Theory of Mind. It is this capacity that allows us to enjoy Shakespeare. We are uniquely capable of taking the basic Golden Rule and morphing it into Game Theory. We are uniquely capable of extending and abstracting empathy, heightening anticipation, extending reward time and producing varied and complex cultures. All animals exhibit the rudiments of these behaviors, but not at the high level that humans do. Sapolsky repeatedly iterates the phrase "we are alone in that realm as well."

After the categories of "Not Unique at All" and "Unique by Degree," Sapolsky goes on to his third and final category, "Simply Unique." This third category represents a different challenge: "Some of the time the challenge is that we are dealing with something where we are simply unique; there is no precedent out there in the animal world." Here, he is pointing to the human ability to sustain a specific kind of cognitive contradiction: "Gaining the strength and will to do X from the irrefutable evidence that X cannot be." For examples of this unprecedented uniqueness, Sapolsky, a self-proclaimed strident atheist, turns first to Danish philosopher and theologian Søren Kierkegaard, and then to Sister Helen Prejean, a Catholic nun. He quotes Kierkegaard on the contradictory nature of faith: "Christian faith requires that faith persists in the face of the impossible, and that humans have the capacity to simultaneously believe in two contradictory things. The very contradictory nature of it is what makes it vital, essential, and a moral imperative" (qtd. in Sapolosky).

The embodiment of this persistence of faith is found in Sister Helen Prejean, a Catholic nun who ministers to death-row inmates in Louisiana and the author of *Dead Man Walking* and *The Death of Innocents*. When she is asked how she can do the work she does, she echoes Kierkegaard in her answer: "The less forgivable the act, the more it must be forgiven" (qtd. in Sapolsky). Sapolsky loves this. He is genuinely moved by the sentiment and the enactment of it: "This strikes me as the most irrational magnificent thing we are capable of as a species. . . . We are the unique-iest simply because of this property of us."

He deftly parlays this into a perfect closure for the occasion of the speech. Part of a college education, he notes, is having one's eyes opened up to the world, a kind of wising up. "And one of the things that happens is that when you've wised up enough, there's a very clear conclusion that you have to reach after a while, which is: at the end of the day, it's really impossible for one person to make a difference." This is the impossibility, he urges, which must be submitted to the logic that "gaining the strength and will to do X [comes] from the irrefutable evidence that X cannot be [done]." Of all the people who are in a position to hold this as a lifelong imperative, he argues, certainly it is a group of Stanford graduates with an education, network and level of privilege that, yes, sets them apart. "So do it," he concludes, and the crowd responds with exuberant applause. They are the best of the best. They are the über unique. The animals have, essentially, told them so.

> The word animal, though it exists as a term of science, does most of its work in areas that are far from being detached and scientific. It serves continually as a reference point in the forming of our communal self-image. . . . In its first, inclusive use it names a class to which we all belong. In its second it names one to which we do not belong, and whose characteristic properties can be used to supply a foil, a dramatic contrast lighting up the human image. (Midgley 36)

WHO FEELS WHAT?

Despite many people's feelings toward modern dentistry, the level of care taken to ensure that the patient is numb to any pain is pretty remarkable. Imagine that you have a cavity, and you go to the dentist. The dentist will need to drill into the tooth and then fill it, but before any of that happens, you are given a shot of Novocain. The dentist gingerly pokes your gums, making sure the anesthetic has kicked in. "Can you feel that?" the dentist asks. If you can feel the poke, you wait a little while until you're numb. The dentist is patient. It is important that you do not suffer pain.

The fate of CH-377 offers a stark contrast both in attitude and practice. CH-377 died of an infection in the New York University Lab. It had been "darted 220 times, once accidentally in the lip. . . . subjected to 28 liver, two bone marrow, and two lymph node biopsies. . . . injected four times with test vaccines, one of them known to be a hepatitis vaccine. . . . injected with 10,000 times the lethal dose of HIV." It died of "an infection aggravated by years of darts, needles, and biopsies" (Bekoff 27).

It was a he. A chimpanzee named Pablo. But Pablo was not among the über unique. Or even the unique-ier. He was just among the mundane unique, in the sense that every species is unique. That's the story that allows for the wild discrepancy in our attitude toward and treatment of bodies— ours and theirs. It has been the story for so long that the majority of people mistake it for reality.

> Medieval and Renaissance theology and philosophy—rooted in the Bible and Aristotle, and confirmed by Descartes, Spinoza and Kant—were wholly anthropocentric: nature was created for the interests of humanity, 'every animal was intended to serve some human purpose, if not practical, then moral or aesthetic.' Man, made in the image of God and endowed with reason, was fundamentally different in kind from other forms of life, which he was entitled to treat as he chose. (Tapper 48)

Georges Bataille, a French writer whose postmortem influence can be found in many philosophical perspectives, railed against the modern lifestyle founded on the simultaneous perpetuation and denial of violence. "'In our time . . . the slaughterhouse is cursed and quarantined like a plague-ridden ship.' . . . 'good folk' are led 'to vegetate as far from the slaughterhouse as possible, to exile themselves, out of propriety, to a flabby world in which nothing fearful remains'" (Bataille qtd. in Bois 46). Today, it's more than the slaughterhouse—it's the Walmart with its sweatshop clothes and below-subsistence wages, it's the genetically engineered produce, the crippled chickens of the fast-food industry and, beyond a doubt, it is the modern animal-testing facilities.

The story that humans are more valuable than other animals persists. We seem unable to celebrate our difference without insisting on a hierarchy in which we are at the top (with Stanford graduates at the top of the top, according to Sapolsky).

For example, at the end of his otherwise excellent book on the enteric nervous system, Michael Gershon, M.D., closes with a discussion of animals in biomedical research. He states that scientists "do not wish to be the agents of the unnecessary death of any sentient being" (Gershon 312). He notes the immense care that is given to animals in these research facilities in terms of living quarters and food. "In contrast to biomedical scientists, disease places no limits on the degree of suffering it causes . . . " (312). These seem to be fair claims. Research animals, after all, cost thousands of dollars apiece, besides which most people are not pathological enough to enjoy tormenting them.

From here, however, Gershon presents some false dilemmas. "People who object to animal research . . . feel for the animals, but they do not feel for people" (312). He continues, "I have never observed moral consistency in animal rights crusaders. . . . If an animal rights person were to renounce the benefits of animal experimentation . . . I would anticipate that he/she would die prematurely and probably painfully" (313). And "Adolf Hitler opposed animal research. . . . His alternative was Dachau" (312).

Scientists—objective, rational scientists who otherwise do very interesting work—do not always realize when the arguments they make to justify their work and their findings are not, as the scientific model would demand, objective or rational, but rooted in systems of thinking, narratives, that tacitly perpetuate an arbitrary set of values:

> In crusading for animal rights, activists often accuse scientists of "speciesism," a term they equate with racism or sexism. Simply to use the term "speciesism," however, is to announce that its user is unable to understand that human life is sacred. Since research on animals alleviates disease and human suffering, demands that this research stop are really proposals to expand disease and promote human suffering . . .
>
> The use of animals to make human lives better has been considered moral throughout history. The human use of animals is found in ancient sources and is explicitly acknowledged in the Bible. "God said: Let us make humankind in our image, according to our likeness! Let them have dominion over the fish of the sea, the fowl of the heavens, animals, all the earth, and all crawling things that crawl about the earth!" This is a mandate that should not be abused by cruelty, but neither should it be forgotten. . . . Attempts by activists to prevent research on animals are harmful to others. . . . Fundamentalism, which extends one's personal beliefs to others, is no better when it is practiced in the name of animals than when it is practiced by Iranian ayatollahs in the name of God. (Gershon 313)

God said: That is Gershon's foundational argument, the story that makes sense of his lifeworld. It is a kind of psychological Novocain.

The reasoning presented by Gershon would not pass muster in a first-year composition or communications course, let alone make it through graduation, medical school and publication. Yet it has. It has because it belongs to the same narrative wherein humans are unique-ier. The values align, the persuasion holds.

But why does is hold so tightly?

The 1988 World Archaeological Congress on 'Cultural Attitudes to Animals' began with a session wherein various contributors tackled the question "What is an animal?" Tim Ingold, the anthropologist who posed the question, wrote this of the session:

> It came as no surprise that my question spawned answers of very different kinds, and that they disagreed on many fundamental points of principle. Perhaps more surprising was the degree of passion aroused in the course of the discussion, which seemed to confirm two points on which I think all the contributors would agree: first, that there is a strong emotional undercurrent to our ideas about animality; and, second, that to subject these ideas to critical scrutiny is to expose highly sensitive and largely unexplored aspects of the understanding of our own humanity. (1)

But these ideas are, perhaps, showing signs of wear. Perhaps it is because we refuse to pay attention to them. Perhaps it is the presence of other narratives and other values, other ideas that emerge like ghost lilies, seemingly overnight, but there they are: Surprise! There is something new in the garden.

A PERSISTENT ILLUSION

> At evening, larkers stalk the wheat fields, nets spread. Bits of mirror flash behind them. Larks fly into the glittering—and the nets. Larkers cage them. Off they go to wealthy tables, waiting mouths, in Turnbridge and Brighthelmstone.
>
> So it is with humans. Quickness draws their eye. Entangles their attention. What they notice they call reality. But reality is a fence with many holes, a net with many tears. I walk through them slowly. My slowness is deceptively fast. (Klinkenborg 18).

Albert Einstein is famously quoted as saying, "Reality is an illusion, albeit a persistent one." As a theoretical physicist, his true north was mystery. For most of the rest of humanity, however, including other scientists, certainty is the preferred narrative currency. But errors in that certainty, while perhaps embarrassing, are wonderful things. They are cracks in the wall, holes in the net, the softening of a border that has too long been solid.

In the "Empathy" part of his lecture on "The Uniqueness of Humans," Dr. Sapolsky shows a slide of a dog, a black lab, with its paw caught in a trap. And apparently it has been there for a while because the paw is beginning to necrose. In commenting upon the empathy that we are feeling for this dog, he remarks that other animals do not do this. "So we are not alone in that [the 'rudiments' of empathy as found in chimps]. Where we are alone is just the extraordinary directions we can take that empathy. . . . we are feeling empathy for a member of another species. This is unheard of."

Well, actually, it's not; it's heard of rather frequently. Even if one discounted the many undocumented experiences of empathy that we humans have with our household pets, consider the following:

> In but one of many stories of dolphins helping humans at sea, in New Zealand, a pod of dolphins circled protectively around a group of swimmers to fend off an attack by a great white shark. "They started to herd us up. They pushed all four of us together by doing tight circles around us," said Rob Howes, one of the swimmers. In these stories, we see that the empathetic presence of animals can have a direct and immediate impact on our well-being and even survival. (Bekoff 17)

There is also the deep bond between Tara the elephant and Bella the dog at the Elephant Sanctuary in Tennessee ("Animal Odd Couple"), the crow who takes care of the kitten ("Crow Adopts Kitten"), the leopard and the golden retriever ("Spot the Difference"), the lions who rescued the little girl from kidnappers (Bekoff 16), the dog who nearly died guarding kittens from a fire ("Leo the Lionhearted"), and on and on and on. This is not a matter of data; it is a matter of paradigm, of the familiar story Sapolsky is telling, the story that keeps everything in place.

After discussing the dog with its paw caught in the trap, he moves on to Picasso's *Guernica*, focusing on the horse in the middle to show how we can empathize at an abstract level with a horse burning to death in its stable as the fascist bombs fall from above—even when that horse is a painting. Continuing in the same vein, Sapolsky shows a slide of Franz Marc's painting *The Fate of the Animals*. He mentions Marc's experience with the trench warfare of World War I and uses this to springboard into further praise of our empathetic capacities:

> And what you have here is sheer utter chaos breaking out all around—no doubt the emotional viscera of what World War I was about. And in the center, an animal baying at the moon in the sort of terror of this chaos, and an animal of a shape, appearance, color that does not even exist on earth. This is a purely imaginary animal. And we sit here, and understanding where this came from and what he had experienced to paint something like this, we're not feeling sorry for that animal. We're feeling sorry for *the animals*, for all the innocent victims—we are taking empathy to a realm that no other species can imagine.

Now, Stanford professors who publish books and conduct fieldwork and lab research as well as teach classes are busy people. In that context, it can be easy to miss something. It's unfortunate that the thing Sapolsky missed about this painting was the date of this painting—1913, the year *before* the war started.

The Fate of the Animals was Marc's response to logging and industrialization. In fact, the original title was *The Trees Show Their Rings, The Animals Their Veins*, and on the back of the painting he had written "And All Being is Flaming Suffering"; he changed the title to *Tierschicksale*, or *The Fate of the Animals*, under the influence of his friend Paul Klee (Levine 269). It is not a response to the war, but more a premonition of it:

> While stationed at the [western] front in the early spring of 1915, the artist received a postcard from his friend and patron Bernhard Koehler. On the front of the card was a reproduction of this painting and its effect on the artist was both immediate and profound. . . . On the same day, Marc addressed a letter to his wife in which he acknowledged receipt of Koehler's card and then went on to say, "At first glance I was completely shaken. It is like a premonition of this war, horrible and gripping. I can hardly believe that I painted it! . . . It is artistically logical to paint such pictures before wars, not as dumb reminiscences afterward . . . " (Levine 269)

To make such a painting is like an animal anticipating an earthquake, an enviable intuition. As poet Dorothy Barresi puts it: "To know something's coming, anything, sub rosa / in the meat and tender / architecture of our paws! / Of course we have no such wild sense, // but what if we did?" (6–7).

And what if we do? What if this is something we have in common with other animals, the ability to sense approaching danger? To act on our unease? Of course, for us to look at that area of commonality would not result in humans getting the highest marks. And that is, after all, the underlying message in Sapolsky's speech: humans are the most unique species on earth.

As stated earlier, this is not a matter of the data, but of paradigm:

> It is *not just* a matter of asking whether one has the right to refuse the animal such and such a power (speech, reason, experience of death, mourning, culture, institutions, technics, clothing, lying, pretense of presence, covering of tracks, gift, laughter, crying, respect, etc.—the list is necessarily without limit, and the most powerful philosophical tradition in which we live has refused the "animal" *all of that*). It *also* means asking whether what calls itself human has the right rigorously to attribute to man, which means therefore to attribute to himself, what he refuses the animal, and whether he can ever possess the *pure, rigorous, indivisible* concept, as such, of that attribution. (Derrida 135)

All of this is not to say that other animals have culture that is as complex as ours or that they make as many and as varied tools, etc. But it is to ask this: (1) To what end do we continually find that our comparisons between humans and animals necessitate the conclusion that we are so significantly different that our difference alone is sufficient means by which to know ourselves as humans? and (2) By what pretense of objectivity or rationality have we determined that those differences make us better than those we have categorized by *our criteria*?

There is no scientific or rational accounting for this; there is only the story.

THE SKIN OF THE WORLD

The present pattern of our collective logic regarding earth's life forms is to articulate a variety of binaries such as unique/ubiquitous, complex/simple, language/silence, individual/group, *human/animal* and then cling desperately to one side. The truth, however, is that we are good and bad, vulnerable and strong, masculine and feminine, divine and animal. Or more accurately, we are between these perceived ends of the spectrum; it is in the play between them where the flow of experience finds us, tipping first one way, then the next. Differences are felt as contrasts, as skin touching skin. The rigidity and expectation of stasis numbs us not only to the world, but also our own potential. It is as if the real agenda is to pin the specimen to the velvet backing of the box.

Peters tackles this very issue in his elegant book *Speaking into the Air: A History of the Idea of Communication*. Granting humanity the full range of our potential as communicators, he winds from the Synoptic Gospels to computers and angels, and he has this to say about interspecies communication:

> The task in animal communication is to find a fellow intelligence in a body of a different shape and species. The task is to find affinities not limited by our anthropomorphic dispositions. . . . "Communication" gives us an image of humanity, not as standing on an ontological ladder betwixt the beasts and the angels, but as a nexus within a biological network and circuit of information flows. (Peters 230–244)

This is a conversation of interstitial spaces, of amongs and betweens, of function, process and, above all, relations. It is about the stories we tell as we move from node to node of reckoned being, and the stories that govern each one when we arrive. It is about the logic of the rhizome as it bends and turns, amasses and shoots off, and, as any decent gardener knows, the need at times to dig up the mass, break it apart and replant it so that it stays healthy. Recognizing that we are at a time to break apart our amassed stories of the human-animal difference may be uncomfortable, but we must not anesthetize ourselves. This is the time to let our skins feel, to do the work of connection, even if it feels strange. Even if it hurts.

> Given our condition as mortals, communication will always remain a problem of power, ethics, and art. Short of some redeemed state of angels or porpoises, there is not release from the discipline of the object in our mutual dealings. This fact is not something to lament: it is the beginning of wisdom. To treat others as we would want to be treated means performing for them in such a way not that the self is authentically represented but that the other is caringly served. This kind of connection beats anything the angels might offer. Joy is found not in the surpassing of touch, but in its fullness. (Peters 269)

IT HAS ALL BEEN SO BEAUTIFUL: A SMALL STORY FOR DERRIDA

In *The World We Have*, Thich Nhat Hanh reminds readers that the world may die, and we need to accept this. To do this, we must accept our own deaths. But mortality, like the slaughterhouse, is too uncomfortable a reality for most. We want the garden to be a place of order, control. Of peace. We want to dwell in the preferred half of our fictional binaries. But the garden of the earth is wildly tangled. Life and death are intertwined more tightly than lovers, and we don't get one without the other. And, most important, one is not better than another. Recognizing that life is precious means recognizing that death is profound. We are able to actually experience so little of this world—what does it serve to cordon ourselves off when time already does so much of that for us?

Time, or the lack thereof, is an underlying theme in Jacques Derrida's small book *The Animal That Therefore I Am*. It was finished after his death in a sutured manner, his community stitching together lecture text, notes, transcripts of discussions. In it, he mourns the lack of time that he and his comrades will have together, but also that so little time has been spent on this important issue of humans and animals. "If I had time and if we had time together. . . . we don't have time. . . . if I had time I would try to show how. . . . we won't have time to go very far. . . . [. . .] one should spend a long time on this. . . . " (Mallet xii). The text bears the marks of its permeability in ellipses and dashes, and there was a little sentence, spoken in 1997, that weaves constantly throughout it, and thus through his readers and friends. A little chime, a thought-bell that is struck to remind us, to bring us into awareness of this world we have. A little sentence that made this text possible, then impossible and, finally, possible in a different way: "Life will have been so short" (Mallet xiii). Perhaps it is the realization of that brevity that fuels the human need to set ourselves apart, to tell stories that make our time feel more important than the time of others. But it could work the other way. The realization that life is brief could just as easily generate stories wherein we learn that everyone's time is precious, everyone's life is important.

Given our present culture, however, these stories do not come easily: Recently, someone walked past my front yard, where I am growing vegetables, flowers, melons and herbs, and took off with a watermelon. The watermelon vines had been tangled and ripped up in some places, and as I set them back into the earth, I realized that *all* the other ripe melons were gone as well. Who would do such a thing? Clearly not someone who gardened, who labored against the weeds, carried water to the thirsty roots and lovingly talked to her plants. I would have been less upset if the young rabbit the cat killed had, instead, made its way through the patch, feasting.

Yet I killed the squash bugs as if their massacre was justified. Are they less wild than a rabbit? Or is it me—am I more wild than the stories of my humanity have led me to believe? And if I am more wild, capable of brutality, what am I to presume about the watermelon thief, about his or her humanity, or hunger? The slaughterhouse and the garden are supposed to be different, one human and the other animal. But these experiences, these encounters reflect myself back to me, revealing the contour of a life that strains at the narrative boundaries of what my human experience is supposed to mean. I meet myself quite often in the garden, where plants and animals, caretakers and thieves seem to converge. Which of us is right? Which of us is best? Our permeability leaves me uncertain and raw. Recognizing these mutually constituted existences, acknowledging the slaughterhouse in the garden—and vice versa—pulls me into the orbit of a different story, a different lifeworld. Difference is everywhere, but nothing is fixed, ranked. Like Derrida's center, it exists, but it does not hold. This narrative paradigm is engaged in a different project than that of the Enlightenment. Knowledge—life—is not something to pin down and examine. It swirls in, around and through us, a kaleidoscopic mirror, and we see ourselves in the gloating thief or the weary chimp or the benevolent dolphins. And we are left echoing Derrida: It is all so beautiful. Laughably brief.

NOTES

1. These oils are powerful octopamine blockers. Octopamine is a neurotransmitter that honeybees use to remember the location of food sources as well as communicate with other bees.
2. This is in contrast to those governing narratives that have, as a metavalue, the *exclusion* of new information so that other facets of their governing narrative are not subject to change. I am here thinking particularly of fundamentalist rhetoric as it occurs across many different religions as well as in conservative politics.
3. Judicious in the sense that those deploying such words have careers to maintain, and such concerns would logically bear on how they present their observations. Jane Goodall learning to apply scientific conventions to her observations is an example.

4. "The concept of narrative rationality asserts that it is not the *individual form* of argument that is ultimately persuasive in discourse. That is important, but values are more persuasive, and they may be expressed in a variety of modes, of which argument is only one" (Fisher 48).

REFERENCES

Barresi, Dorothy. *The Post-Rapture Diner*. Pittsburg: U of Pittsburg P, 1996. Print.

Bekoff, Marc. *The Emotional Lives of Animals*. Novato, CA: New World, 2007. Print.

Bird, Alexander. "Thomas Kuhn." *Stanford Encyclopedia of Philosophy*. Stanford U, 4 Aug. 2004. Web. 17 Aug. 2010.

Bois, Yves-Alain, and Rosalind E. Kraus. *Formless: A User's Guide*. New York: Zone, 1997. Print.

Conquergood, Dwight. "Performance Studies: Interventions and Radical Research." *The Drama Review* 46.2 (2002): 145–156. Print.

"Crow Adopts Kitten." *Animal Planet*. 2 Nov. 2006. Web. 10 Aug. 2010. < http:// wn.com/kitten_makes_friends_with_a_crow!!>.

Derrida, Jacques. *The Animal That Therefore I Am*. Trans. David Wills. Ed. Marie-Louise Mallet. New York: Fordham UP, 2008. Print.

Dickinson, Emily. "A Word Is Dead." *The Complete Poems of Emily Dickinson*. Ed. Thomas H. Johnson. New York: Back Bay, 1961. Print.

Fisher, Walter R. *Human Communication as Narration: Toward a Philosophy of Reason, Value, and Action*. Columbia, SC: U of Southern California P, 1987. Print.

Goodall, Jane. Forward. *The Emotional Lives of Animals*. By Marc Bekoff. Novato, CA: New World, 2007. Print.

Jandér, Charlotte, and Edward Allen Herre. "Host Sanctions and Pollinator Cheating in the Fig Tree—Fig Wasp Mutualism." *Proceedings of the Royal Society B: Biological Sciences* 277 (2010): 1481–1488. Web. 16 Aug. 2010. <http://striweb. si.edu/publications/PDFs/STRI-W_Herre_2010_Host_Sanctions_NV.pdf>

Klinkenborg, Verlyn. *Timothy; Or, Notes of an Abject Reptile*. New York: Knopf, 2006. Print.

Kuhn, Thomas. *The Structure of Scientific Revolutions*. Chicago: U of Chicago P, 1996. Print.

"Leo the Lionhearted." *World News*. 27 Oct. 2008. Web. 10 Aug. 2010. < http:// www.examiner.com/article/leo-the-lionhearted-dog-refuses-to-leave-kittens-burning-house-video>.

Levine, Frederick S. "The Iconography of Franz Marc's Fate of the Animals." *The Art Bulletin* 58.2 (1976): 269–277. *JSTOR*. Web. 14 Aug. 2010.

Long, Cheryl. "Squish Those Squash Bugs."*Mother Earth News* Aug.–Sep. 2010. Print.

Mallet, Marie-Louise. Forward. *The Animal That Therefore I Am*. By Jacques Derrida. Trans. David Wells. Ed. Marie-Louise Mallet. New York: Fordham UP, 2008. Print.

Midgley, Mary. "Beasts, Brutes, and Monsters." *What Is an Animal?* Ed. Tim Ingold. New York: Routledge, 1988. 35–46. Print.

Nhat Hanh, Thich. *The World We Have: A Buddhist Approach to Peace and Ecology*. Berkeley, CA: Parallax, 2008. Print.

Oliver, Mary. "Snake." *House of Light*. Boston: Beacon, 1990. Print.

Peters, John Durham. *Speaking into the Air: A History of the Idea of Communication*. Chicago: U of Chicago P, 1999. Print.

Ramanujan, Krishna. "Study: Trees Retaliate When Fig Wasps Don't Pollinate Them." *Cornell Chronicle Online* 27 Jan. 2010. N. pag. Web. 15 Aug. 2010.

Sapolsky, Robert. "Class Day Lecture 2009: The Uniqueness of Humans." *You-Tube*. Web. 13 Aug. 2010.

"Spot the Difference: The Leopard and Golden Retriever Who Are the Best of Friends." *Mail Online*. 6 May 2010. Web. 10 Aug. 2010. <http://www.daily-mail.co.uk/news/article-1273864/Salati-leopard-Tommy-dog-snuggle-daily-game-chase.html>

Tapper, Richard. "Animality, Humanity, Morality, Society." *What Is an Animal?* Ed. Tim Ingold. New York: Routledge, 1988. 47–60. Print.

"The Animal Odd Couple." *Assignment America*. 6 Jan. 2009. Web. 10 Aug.2010. < http://www.cbsnews.com/video/watch/?id=4696315n>.

Contributors

Joseph Abisaid (PhD, University of Wisconsin-Madison, 2012) is visiting assistant professor at the University at Albany-SUNY, and teaches courses in communication theory, interpersonal communication, media effects and persuasion. His current research interests focus on media framing and animal rights issues in communication.

Tony E. Adams (PhD, University of South Florida, 2008) is assistant professor in the Department of Communication, Media and Theatre at Northeastern Illinois University. He is the author of *Narrating the Closet: An Autoethnography of Same-Sex Attraction* (Left Coast Press, 2011) and is coediting *The Handbook of Autoethnography* (Left Coast Press) with Stacy Holman Jones and Carolyn Ellis.

Leigh A. Bernacchi is a postdoctoral social scientist with the transdisciplinary Regional Approaches to Climate Change—Pacific Northwest Agriculture program at the University of Idaho and a doctoral candidate at Texas A&M University. Her research focuses on public participation and natural resource management oriented toward endangered species. Her interests ripple from environmental communication, ranging from nature writing, photography and wilderness activism to cultural critical studies. She is an active birder.

Deborah Cox Callister is a doctoral candidate in communication at the University of Utah. Her research and teaching interests include rhetoric, environmental communication and conflict studies with an emphasis on dialogue, collaborative learning and internal coalition communication. She holds a BA in biological sciences from the University of California at Davis and an MS from the University of Utah. She is coauthor of a chapter on rhetorical framing in *Social Movement to Address Climate Change*.

Carrie Packwood Freeman (PhD, University of Oregon, 2008) is assistant professor of communication at Georgia State University in Atlanta. As a media researcher, she studies strategic communication for social change, media ethics, environmental communication and critical animal studies, specializing in animal agribusiness and veganism. She's been active in the animal rights and vegetarian movement for two decades and currently cohosts weekly animal rights and environmental programs on indie radio WRFG-Atlanta.

Susan Hafen (PhD, Ohio University, 1995) is professor of communication at Weber State University. She teaches organizational communication, communication theory, interpersonal and small-group communication and interviewing. Her previous publications have focused on workplace gossip, diversity and critical pedagogy.

Shana Heinricy is a doctoral candidate in the Department of Communication and Culture at Indiana University in Bloomington. Her research and teaching areas include rhetoric, feminist theory and media studies. Her research centers on how mediated and nonmediated human and animal bodies interact with technology. She is particularly interested in how bodies animate discourses of citizenship.

Susannah Bunny LeBaron is a doctoral candidate in speech communication at Southern Illinois University. Her research areas include environmental communication and sustainability, performance studies, narratology and breath training as transformative praxis. She has published several pieces of creative writing, coauthored two thesauri for Houghton-Mifflin and presented numerous performances dealing with the Buddhist idea of interdependent coarising, particularly between humans and other animals.

Stephen J. Lind is a doctoral candidate in the transdisciplinary Rhetorics, Communication and Information Design program at Clemson University. His teaching and research interests are diverse, including argumentation, animal communication, historical puppetry, digital media creation and portrayals of religion in entertainment media. His dissertation explores media representations of religious belief and action through a case study of the *Peanuts* franchise.

Tema Milstein (PhD, University of Washington, 2007) is assistant professor at the University of New Mexico. Her primary research and teaching focus is on critical and interpretive inquiry into cultural perceptions, practices and productions of nature. She is a 2012 Fulbright Scholar and serves on the editorial board of *Environmental Communication: A Journal of Nature and Culture.*

Pat Munday (PhD, Cornell University, 1990) is professor and department head of technical communication at Montana Tech. His teaching and research include science and technology studies, semiotics and environmental studies with an emphasis on citizen participation in policy decisions. His publications range from the history of science to a geographical history of the Big Hole River and its people. He was recently a Fulbright Scholar with Southwest University in Chongqing, China.

Mary Pilgram (PhD, University of Kansas, 2006) is associate professor of communication at Washburn University. Her teaching and research areas include organizational and interpersonal communication with specific interests in training and development, sexual harassment and social support in the areas of pet loss and grief. She has published in both applied and academic venues and presented her research at regional and national conferences.

Emily Plec (PhD, University of Utah, 2002) is professor and chair of communication studies at Western Oregon University. Her research and teaching interests include rhetoric, media studies, intercultural communication and environmental communication. She has authored articles on topics ranging from racism in sports to the rhetoric of women leaders and migrant farmworkers. Her work has appeared in *Environmental Communication: A Journal of Nature and Culture* and other outlets.

Nick Trujillo (PhD, University of Utah, 1983) was professor of communication studies at CSUS. He published over forty scholarly and trade articles as well as four books, including *Organizational Life on Television* (with Leah Vande Berg), *The Meaning of Nolan Ryan, In Search of Naunny's Grave* and *Cancer and Death: A Love Story in Two Voices* (with Leah Vande Berg). He was also a songwriter and independent music publisher. His alter ego, Gory Bateson, released two solo albums, *Is That Viral Enuf 4 U?* and *That Is (Still) the Question.* He passed away on October 29, 2012, shortly before this volume was published.

Index

Made in United States
Orlando, FL
05 February 2022

14493216R00161